The Atlas of COASTS & OCEANS

Ecosystems, Threatened Resources, Marine Conservation

The Atlas of COASTS & OCEANS

Ecosystems, Threatened Resources, Marine Conservation

Don Hinrichsen

The University of Chicago Press
Chicago

The University of Chicago Press,
Chicago 60637

A CIP record for this book is available
from the Library of Congress.

ISBN-13: 978-0-226-34225-2 (cloth)
ISBN-10: 0-226-34225-5 (cloth)
ISBN-13: 978-0-226-34226-9 (paper)
ISBN-10: 0-226-34226-3 (paper)

Produced for the University of Chicago Press by
Myriad Editions
59 Lansdowne Place
Brighton BN3 1FL, UK
www.MyriadEditions.com

Printed on acid-free paper produced from sustainable sources.
Printed and bound in Hong Kong through Lion Production
under the supervision of Bob Cassels, The Hanway Press, London

20 19 18 17 16 15 14 13 12 11
1 2 3 4 5

Edited and coordinated by Dawn Sackett
with Jannet King and Candida Lacey
Designed by Isabelle Lewis and Corinne Pearlman
Maps and graphics created by Isabelle Lewis

Cover photo credits:
Front cover, main image: Dan Barnes/iStockphoto; small images, left to right: NASA; Eugene Suslo/iStockphoto; Pawel Borowka/iStockphoto; Andrew Reese/iStockphoto. Back cover, main image: Davor Lovincic/iStockphoto; small images, left to right: Don Hinrichsen; Andrew Dorey/iStockphoto; Klaas Lingbeek-van Kranen/iStockphoto.

CONTENTS

PART FIVE SEAS IN CONFLICT

PART SIX MANAGEMENT OF COASTAL AND MARINE AREAS

Photo Credits

We would like to thank the following photographers and organizations who have supplied images: page 22–23: Isabelle Lewis; 24: Don Hinrichsen; 29: Don Hinrichsen; 31: Mike Norton/iStockphoto; 32–33: Inga Nielsen/iStockphoto; 34: Klaas Lingbeek-van Kranen/iStockphoto; 36: NASA; 39 (top to bottom): Rhoberazzi/iStockphoto, Kaido Kärner/iStockphoto, NASA; 43: Lesley Jacques/iStockphoto; 45: microgen/iStockphoto; 46–47 (left to right): Davor Lovincic/iStockphoto; Zeynep Mufti/iStockphoto; 48–49 (top left to right) Piero Malaer/iStockphoto, Pawel Borowka/iStockphoto, Lee Chin Yong/iStockphoto, Mark Evans/iStockphoto, Dirk-Jan Mattaar; 52–53: Isabelle Lewis; 55: Linda Steward/iStockphoto; 58: Eugene Suslo/iStockphoto; 65 (top to bottom): Jean Gill/iStockphoto, Zeynep Mufti/iStockphoto, tella_db/iStockphoto; 66: Don Hinrichsen; 68–69 (left to right): Nicky Gordon/iStockphoto, Don Hinrichsen; 70–71: Antonio Scarpi/iStockphoto; 73 (top to bottom): Øystein Lund Andersen/iStockphoto, Don Hinrichsen; 76–77 (left to right): Wolfgang Steiner/iStockphoto, Mark Lynas; 80: Todd Winner/iStockphoto; 82:Michel de Nijs/iStockphoto; 85 (left to right): National Snow and Ice Data Center, Mogens Trolle/iStockphoto; 86–87: Brasil2/iStockphoto; 92–93: ricardoazoury/iStockphoto; 94: Olaf Simon/iStockphoto; 101: K. Schafer/Minden Pictures/FLPA; 102: Dirk-Jan Mattaar/iStockphoto; 104 (top to bottom left): Rusty Elliott/iStockphoto, Dave Long/iStockphoto, oriredmouse/iStockphoto, sweetlifephotos/iStockphoto; 105 (clockwise from polar bear to seagrasses): James Richey/iStockphoto, Cornelis Opstal/iStockphoto, Mark Lynas, Jonathon Heaney/iStockphoto, Nikontiger/iStockphoto, Piero Malaer/iStockphoto, Sharon Metson/iStockphoto.

We have made every effort to obtain permission for the use of copyright material. If there are any omissions, we apologize and shall be pleased to make appropriate acknowledgement in any future edition.

About the Author

Don Hinrichsen is an award winning writer and editor, based in London, UK. He has written five previous books, two of which deal with the state of the world's coasts and oceans – *Our Common Seas* and *Coastal Waters of the World, Trends, Threats and Strategies*. He has authored or co-authored four major reports for John Hopkins University's Center for Communication Programs – on food, freshwater, global environment, and urbanization.

He is currently the senior development manager for the Institute for War and Peace Reporting (IWPR) in London and writes frequently on environment, population, and resource issues for a variety of publications in the US and Europe.

He was formerly Editor-in-Chief of Ambio, the journal of the human environment published by the Royal Swedish Academy of Sciences, and former Editor-in-Chief of the World Resources Report, published by the World Resources Institute in collaboration with the World Bank, UNDP, and UNEP.

FOREWORD

"We still like to go beachcombing, returning for the moment to primitive act and mood. When all the lands will be filled with people and machine, perhaps the last need and observance of man will be, as it was in the beginning, to come down and experience the sea."
CARL SAUER (1962)

The *Atlas of Coasts and Oceans* is a very commendable portrayal of the global connections between the world's oceans and seas, their adjoining coasts and large scale, long-term atmospheric and climatic systems. The oceans and seas are immense, covering 70 percent of the planet. Most of the numerous atlases of the worlds' oceans and seas to date have consisted of global and hemispheric maps, intended to inform readers of various aspects of the marine world – such as the bathymetry, boundaries, resources, and currents. The *Atlas of Coasts and Oceans* complements, rather than replaces such publications. It is not another attempt to map the seas and their coastlines, but rather an attempt to map human impacts upon them. To explore their richness, both intrinsic and in terms of the value humanity accrues from them and to lay out recent management attempts that flow from our realization that the oceans are not a limitlessly renewable resource, but one in need of protection.

A key point that sounds throughout the Atlas is the management implementation gap. This is a universal phenomenon that cuts across all public policy areas. It is certainly not confined to Integrated Coastal and Ocean Management (ICOM). The gap is between completing a management plan/program and providing the powers, creating appropriate institutions, allocating the budget, and retaining competent staff to implement it. It appears that most – if not the great majority of ICOM efforts – have not been able to bridge that gap, particularly those efforts in developing nations and nations with a rapidly emerging economy. The root of problem is that there are very few rigorous and unbiased assessments of ICOM efforts to determine what has worked, what did not work, and why. For the most part ICOM is not learning from the experience of its 45 year history.

The Atlas cites the global databases of ICOM that I compiled between 2000 and 2002. Of the 166 different efforts in 145 coastal countries and semi-sovereign states that I cataloged, how many still exist? And of those, how many have resulted in significant on-the-ground achievements? The failures will largely be in developing nations where the international assistance community kept the effort alive during the plan preparation phase yet generally withdrew the funding required for the long-term process of implementation. Further, undermining the support required for effectiveness and undercutting funding, the international assistance community, by-and-large, no longer regards ICOM as an effective and efficient means to achieve sustainable development goals.

International ICOM efforts are very vulnerable to the implementation gap because participation is voluntary. The Atlas highlights two such efforts: UNEP's Regional Seas Program, and the Large Marine Ecosystems Program. Both have had far more failures than successes. If the international community cannot provide the resources to eradicate piracy in the Red Sea, the Gulf of Aden, and the western Indian Ocean, what can be expected from them in respect of the international management of coastal and ocean resources and environments?

With few exceptions, ICOM has never been supported by a large and cohesive constituency which shares mutual goals. Shipping, offshore oil and gas production, and port development interests are not usually concerned about the declining harvest of capture fisheries and loss or degradation of productive coastal environments. The capture fisheries constituency is weakened by conflict on the management of different stocks and the employment of different technologies. There are further conflicts between capture fisheries and mariculture farmers.

By showing the richness of the seas and their coastlines, and illustrating human interactions with these rich environments, the general reader gains a deeper understanding of just how important management efforts are, and how much we stand to lose if they don't succeed.

On our increasingly developed shores has beachcombing for many – if not most – people become more of a participation in a periodic litter picking event than an idle stroll and the opportunity to chance upon weathered treasures?

Jens Sorensen,
Environmental Policy and Planning,
Integrated Coastal Management

Our Ocean Planet In Peril

Humanity has, for generations, depended on the oceans while treating them with a cavalier disregard that has resulted in seriously damaged ecosystems. Healthy oceans are essential to a healthy terrestrial environment and many benefits from the oceans that we have taken for granted are now threatened by our actions. The oceans are not yet fully explored, let alone understood and new species continue to be identified every year. Given our dependence on the world's seas, we underestimate their complexity and importance at our own peril.

Burgeoning human numbers and growing consumption are putting intense pressure on ocean and coastal areas, undermining the health of the oceans themselves and leading to depletion of ocean resources.

Human populations have a tremendous impact on the quality of coastal and oceanic environments. Roughly 40 percent of the world's population – around 3 billion people – occupy a coastal strip 100 kilometers (60 miles) wide, representing only 5 percent of the earth's land surface. The resulting development pressures are taking a grim toll on coastal and near-shore resources.

Of Asia's total population of 4.1 billion, over one billion live within 100 kilometers (62 miles) of the sea. The exceptions are India, Pakistan, and, of course, the land-locked countries of Central Asia. The population of Latin America and the Caribbean is even more clustered on the coasts. The region's coastal states had a total population of 569 million (in 2009); a full three-quarters of them living within 200 kilometers (120 miles) of a coast.

Human actions are currently creating unprecedented carbon dioxide emissions, leading to global warming. The ocean has acted as an important carbon sink, sequestering a high proportion of this carbon dioxide and thereby moderating human impacts on global climate. The cost is an increasingly acidified ocean. This is having a direct impact on marine organisms, the so-called marine calcifiers, which use calcium to build their protective shells. Increasing acidification prevents these organisms from functioning properly and impedes their ability to build and repair their shells. This is leading to declines in corals, shellfish and some plankton populations, among others, which leads to loss of habitats for fish and reduced foodstocks for the marine food chain.

Global climate change is also warming the seas, causing coral bleaching, creating new dead zones, and contributing to sea-level rise (the seas warm as they expand, thus taking up more room). The melting of land ice on Greenland and Antarctica will add to the trend towards rising seas. Rising seas, in turn, could have devastating impacts on coral reefs and seagrass beds, among other shallow coastal, near-shore and ocean habitats by causing more turbidity in the water column and by raising the water levels beyond which these ecosystems can survive

Critical Ocean and Coastal Ecosystems Facing Collapse

Scientists have mapped the impact of 17 anthropogenic drivers of ecosystem damage and change across 20 vital marine ecosystems, including coastal wetlands, coral reefs, seagrasses, and seamounts, among others. Their conclusion: approximately 41 percent of the world ocean is suffering a moderately-heavy to heavy impact either directly, from human activities (e.g. over-fishing and habitat destruction) or indirectly, from human induced effects (such as climate change). Only four percent of the ocean remains relatively unaffected by human activities. The worst hot spots in terms of degraded habitats and persistent pollution include: the North Sea, South and East China Seas, the East Coast of North America, the Caribbean, the Mediterranean and Red Seas, the Bering Sea, part of the western Pacific around Japan, and the Persian Gulf.

The picture is indeed grim. The world has lost one quarter of its salt marshes, 30 percent of its mangrove forests and one-third of its seagrass meadows over the past 60 years. We are losing these fecund coastal ecosystems two to fifteen times faster than the rate of tropical forest loss.

Mangrove habitats shelter over 2,000 species of fish, shellfish, invertebrates and plants, but in the Philippines, for instance, the mangrove area has been reduced by 90 percent, from one million hectares in 1960 to around 100,000 hectares by 2000. The mangroves have been replaced by shrimp and fish ponds, or clear cut to make way for agricultural development. Worldwide, the mangrove area has decreased by 3.6 million hectares since 1980.

Seagrass beds (e.g. eel and turtle grass), kelp, seaweeds, and *Posidonia* colonies – the underwater meadows of the ocean – have fared little better. They have declined by one-third over the past six decades. With a few exceptions, these diverse ecosystems appear to be in retreat near virtually all inhabited coastal areas.

Coral reefs, the rainforests of the sea, are being destroyed in the name of development. Of the world's 285,000 square kilometers (110,039 square miles) of reefs found in 100 tropical and semi-tropical countries, scientists estimate that, as of 2008, over half are either degraded beyond recognition, in critical condition, or threatened. In South-East Asia, for example, one of the epicenters of coral biodiversity, close to 90 percent of all reefs are at risk. However, the remaining 46 percent could very well be lost by mid-century, killed off by climate change (increasing water temperatures, acidification, and coral bleaching), over-exploitation, including reef-based fisheries, and pollution, among other impacts

Pillaging the Seas

Coastal and ocean fisheries – the largest harvest of a wild food source on the planet – are in serious trouble. Commercial fishing fleets landed around 81.9 million metric tons of seafood in 2006, according to the Food and Agriculture Organization (FAO). In 2007, nearly 20 percent of fisheries were over-exploited, 52 percent were fully exploited and 18 percent were moderately exploited. Only two percent of stocks were considered under-exploited.

Productivity has fallen over the past few years in all but four of the world's 15 major fishing regions, as defined by FAO. Landings of the most valuable species of fish, including cod, tuna, and haddock, have dropped by one-quarter since 1970. In the four hardest hit regions – the northwest, the west-central and southeast Atlantic, and the east-central Pacific – catches have plummeted by more than 30 percent since 1989.

The demise of commercial fisheries in nearly every sea is a classic example of the tragedy of the commons – since no one owns the resource, everyone exploits it to the point of collapse. Moreover, without agreed international management plans in place and actually enforced, there is little hope that wild stocks will ever recover.

Take the example of the Black Sea. Once blessed with nearly 30 species of commercially valuable fish, this nearly enclosed sea saw its catches plunge by some 900,000 tons over the 30-year period from 1960 to 1990. In many areas (except in Turkish waters) the sea no longer has much in the way of exploitable marine life. The Black Sea is well on its way to becoming a dead sea in terms of biological diversity due to pollution inputs over the past four decades and rampant over-fishing. All waters below 150 to 200 meters are virtually without oxygen, a natural phenomenon made worse by the pollution ferried in by its vast watershed which drains the eastern half of Europe. Only 10 percent of the total volume of near-surface water has enough oxygen to sustain marine life higher than micro-organisms.

The Black Sea is also a study in invasive species. The comb jelly, *Mnemiopsis leidyi*, imported from the Atlantic, has thrived in its fish-scarce waters. In 2005, a study found that this voracious jellyfish, lacking predators, comprised 90 percent of the sea's biomass.

The disappearance of the world's marine catch of fish and shellfish has ominous implications for the food supply of the 2.9 billion people who depend on the sea for at least 15 percent of their daily protein intake. Seafood provides close to 20 percent of the world's total food supply. In South-East Asia and the South Pacific, the sea provides up to 100 percent of all animal protein in daily diets.

The world's fishing fleets discard at least 20 million metric tons of fish and shellfish every year as by-catch from their operations. Most of the waste is due to trawlers which harvest enormous quantities of marine life in their relentless search for squid, shrimp or bottom dwelling fish (such as halibut, sole, and flounder). This loss of potential protein (and income) amounts to nearly one-quarter of the world's annual take from the seas. Losses from discards in the Bering Sea and Gulf of Alaska alone have been estimated at around $250 million a year.

Can mariculture and aquaculture substitute for the declining ocean fish catch? Enormous increases in farmed seafood, especially in East Asia, are making up for much of the shortfall in wild catches. In 2007 the FAO reported that 51 million metric tons of fish, shellfish, and seaweeds were farmed (both marine and freshwater), worth $86 billion. Aquaculture and mariculture operations have grown by over 8 percent per year over the past decade. China accounts for the bulk of this by volume, producing close to 70 percent of all farmed species. However, much of China's aquaculture production is sent to cities to feed rising demand, or exported to industrialized countries, such as the table-ready shrimp from Ecuador or South-East Asia sent to North America and Europe.

Rising Tide of Pollution

The great world ocean, which is at the heart of the global hydrological cycle, is under stress from human activities on land and at sea. Expanding human populations and the growth of cities along coastlines has contributed to a rising tide of pollution in nearly all of the world's seas. Coastal urban areas dump increasing loads of untreated or partially treated industrial and municipal wastes into the sea. In fact, waters around many coastal cities have turned into virtual cesspools, so thick with pollution that virtually no marine life can survive. At sea, ships discharge oily ballast waters and other wastes directly into the water.

The result is a toxic cocktail of untreated sewage and municipal wastes, chemical pollution from industries and oxygen-depleting agricultural chemicals, such as nitrogen and phosphorus pulled off farmlands and washed into coastal waters by rivers and streams. Consistent pollution over time has resulted in the proliferation of dead zones. As of 2008, over 400 of these lifeless areas had been identified, most of them in the waters of developed countries and many in previously prime fishing grounds. Dead zones are areas devoid of oxygen and are, for the most part, incapable of supporting marine life higher than micro-organisms. Human activities, coupled with the inability to manage coastal and ocean areas in an integrated and sustainable fashion, have triggered the degradation and collapse of key ocean and coastal resources.

Consider the following:

- Despite over three decades of cleanup efforts, the Mediterranean Sea is on the receiving end of some 10 billion tons of industrial and municipal wastes each year, much of it with little or no treatment. A full 40 percent of the region's municipalities, containing 14 million people, have no waste water treatment facilities.
- The Lagoon of Iddo in Lagos, Nigeria, receives 60 million liters of raw sewage each year, along with vast quantities of industrial wastes.

- In 2009, the Yangtze River, China's largest, dumped 2.2 million tons of chemical pollutants, as measured by COD, into its estuary, along with 194,000 tons of inorganic nitrogen and 82,000 tons of phosphate.

Ocean currents transport pollutants into the remotest corners of the world's seas. No place in the world ocean is immune from the depredations of humanity. Toxic chemicals, such as PCBs and DDT, for instance, have turned up in the fatty tissues and blubber of seals in the Arctic and penguins in the Antarctic, thousands of kilometers from population centers.

Towards Sustainable Management of Oceans

Coastal and ocean areas can be managed sustainably to preserve key habitats, conserve species and manage fisheries on a sustainable basis. However, this is only possible if governments, civil society organizations join forces with major stakeholders in concerted efforts to actually manage ocean resources instead of simply exploiting the commons.

At present government inaction on ocean management and an inability to enforce existing coastal regulations make problems of overuse, pollution, and resource degradation worse. As of 2002, 145 of the 187 nations, territories and semi-sovereign states with a coastline, had launched integrated coastal management programs. While this number has grown considerably since the 1990s, most countries have yet to move from planning to implementation.

Why is management of coastal and ocean resources so difficult? Coastal areas contain many different jurisdictions – local, regional and national – and involve many different stakeholders. In Brazil, for instance, coastal zone planners have to consult with 20 levels of government. In the United Kingdom, 48 sub-national units of government, from Parliament to town councils, have authority to create an autonomous or semi-autonomous coastal management strategy. Such fragmentation of responsibility makes effective planning and program coordination an enormous challenge.

Nevertheless, there are compelling economic reasons to manage coastal and ocean waters well. Ocean ecosystems provide goods and services worth at least $21 trillion a year, over half of this from coastal ecosystems. The haul of seafood alone is valued at around $80 billion a year and provides direct employment to some 200 million small-scale and commercial fishers. In addition, as many as half a billion people draw their livelihoods indirectly from the sea: processors, packers, shippers, and distributors of seafood; shipbuilders and outfitters; and those working in marine-based tourism and the recreational fishing industry, among others.

There are also vital reasons relating to the ecological value of oceans. For example, coral reefs have been valued at $47,000 per square foot for their shore protection functions alone. In Puget Sound, Washington, just one-third of a hectare of eelgrass is valued at over $400,000 annually in terms of energy derived and nutrition generated for oyster culture, fisheries, and waterfowl.

How can we better manage coastal and ocean resources? The blueprint for a sustainable management system has been outlined by countless national and international organizations over the past three decades. Most require at least eight pre-requisites:

- Local government commitment to project implementation;
- Management awareness of trans-boundary issues, including pollution loads, critical habitats, and endangered species;
- Sound partnerships between implementing agencies;

- Familiarity of various government departments with implementation requirements;
- Indigenous knowledge and expertise to be utilized whenever possible;
- Solid working relationships between government and research/academic institutions;
- Management boundaries to be defined in terms of ecosystems and key features, not geo-political boundaries;
- Research geared to addressing management issues.

The foundation for sustainable management has already been put in place with the coming into force of the Law of the Sea Convention in 1994. It affords all states the right to manage marine resources within their 200-nautical mile Exclusive Economic Zones (EEZs). The problem is, most developing countries do not have the money or the manpower to enforce regulations over such a vast expanse of sea. Small islands in the South Pacific, for example, are dwarfed by their EEZs, which are often 1,000 times larger than the islands which have to manage them.

Though a number of encouraging initiatives have been launched, for the most part they suffer from a lack of implementation and actual enforcement mechanisms. For instance the FAO put in place two well intentioned programs: the FAO Code of Conduct for Responsible Fisheries and the UN Agreement on Straddling Fish Stocks and Highly Migratory Fish Stocks, both launched in 1995/96. All members of FAO have agreed to abide by the Code of Conduct and the UN Agreement on Fish Stocks finally entered into force in December 2001, when Malta ratified it. Sadly, both initiatives suffer from a distinct lack of enforcement mechanisms.

The United Nations Environment Program's regional seas initiative was a sound idea when launched in 1975, with the advent of the Mediterranean Action Plan. Since then, UNEP has launched 12 more regional seas plans. Unfortunately, a review of them in 2010, revealed that only a handful were actually in operation, with a secretariat, a budget, personnel and management capacity on the ground both nationally and regionally.

Integrated coastal zone management is a complicated and multi-agency endeavor that involves coordination of both coastal and sea areas in a comprehensive and coordinated fashion. Unfortunately, most countries, even highly developed ones, have difficulties in converting drawing board plans into operational action plans, with implementation capacities at all levels of governance.

In the final analysis, governments must take the lead in managing their own waters, cooperating as much as possible with neighboring states through international programs like the Mediterranean Action Plan, East Asia's PEMSEA initiative involving 12 states, the Convention for the Protection of the Marine Environment of the Northeast Atlantic and other regional management efforts. It is not too late to start preserving the ultimate source of all life on the "blue planet".

Don Hinrichsen
London, December 2010

ACKNOWLEDGEMENTS

Acknowledgements are often under-rated and un-read. This is unfortunate, as most writers, including this one, would be lost without a considerable amount of expert assistance, everything from reviewing the science and critiquing individual chapters to editing and honing the text to improve its clarity and readability.

I would like to thank in particular Myriad Editions, which took on the chore of producing this book, providing invaluable guidance and encouragement throughout the spawning process. In particular I want to cite Candida Lacey, without whose dedicated advocacy on behalf of the book it would not have progressed beyond the idea stage. To my long-suffering editors, Dawn Sackett and Jannet King, I want to pay particular note; without them this book would be a much poorer version. And thanks to the superb design work by Isabelle Lewis and Corinne Pearlman it is a pleasure to read.

Of course, a number of people reviewed the various drafts for scientific accuracy. Here I am deeply indebted to two friends of long standing, Jens Sorensen and Stephen Olsen. Both of these scientists also reviewed my last book on coastal and ocean areas published by Island Press. The fact that, after one experience of reviewing and fact checking my prose, they agreed to repeat the exercise, is testimony to their dedication. Both contributed their expertise and considerable knowledge to this project, greatly improving the quality of the analysis in the process.

I also want to thank Ben Halpern, from the National Center for Ecological Analysis and Synthesis, for his review of the Chapter, Marine Ecosystems under Threat, based on his massive study published in Science Magazine in February 2008 (Ben was the principal author, joined by 19 other experts).

Last but not least, I want to thank the World Conservation Monitoring Centre, based in Cambridge, UK, for allowing us to use their excellent maps for mangroves, seagrasses and cold corals, and giving us access to their rich data base.

Despite expert advice and countless reviews, any mistakes in this book are entirely my own.

Don Hinrichsen

Definition of Key Terms

Accretion Build up or accumulation of sediment.

Acidification The ongoing decrease in ocean pH caused by human CO_2 emissions, such as the burning of fossil fuels. The oceans currently absorb approximately half of the CO_2 produced. However, when CO_2 dissolves in seawater it forms carbonic acid and as more CO_2 is taken up by the ocean's surface, the pH decreases, moving towards a less alkaline and therefore more acidic state.

Anoxic Devoid of oxygen.

Anthropogenic Resulting from or produced by human activity.

Aquaculture The farming of aquatic organisms including fish, molluscs, crustaceans, and aquatic plants with some sort of intervention in the rearing process to enhance production, such as regular stocking, feeding, protection from predators, etc.

Biological Oxygen Demand (BOD) A measure of the oxygen used by microorganisms to decompose organic wastes such as raw sewage, as well as nitrates and phosphates from farmland.

By-catch Part of a catch of a fishing unit taken incidentally in addition to the target species towards which fishing effort is directed. Some or all of it may be returned to the sea as discards, usually dead or dying.

Carbon dioxide (CO_2) A naturally occurring gas, also a by-product of burning fossil fuels and biomass, as well as land-use changes and other industrial processes. It is the principal anthropogenic greenhouse gas.

Carbon sinks Reservoirs for carbon, such as forests and oceans, that sequester more carbon than they release.

Chemical Oxygen Demand (COD) A measure the oxygen used by microorganisms to break down organic and inorganic pollutants, mostly consisting of industrial wastes.

Climate Change A statistically significant variation in either the mean state of the climate or in its variability, persisting for an extended period (typically decades or longer). Climate change may be due to natural internal processes, or to persistent anthropogenic (human) changes in the composition of the atmosphere or in land use. The UNFCC its Article 1 defines it as: "a change of climate which is attributed directly or indirectly to human activity that alters the composition of the global atmosphere and which is in addition to natural climate variability observed over comparable time periods." They thus distinguish between "climate change" attributable to human activities and "climate variability" attributable to natural causes.

Coral bleaching The paling in colour of corals resulting from the loss of symbiotic algae, in response to abrupt changes in temperature, salinity and/or turbidity.

Dead zones Regions of the ocean where dissolved oxygen has fallen to such low levels that most marine species can no longer survive. Such conditions are often seasonal. They are mostly caused by agricultural runoff, especially nitrogen-rich fertilizers, as well as the burning of fossil fuels. Pollutants from these sources cause marine eutrophication, whereby the ecosystem receives too many nutrients, triggering massive algal blooms, which eventually die and are broken down by bacteria. In the process the bacteria consume excessive amounts of oxygen, essentially starving the marine system.

Drivers A driver is any natural or human-induced factor that directly or indirectly causes a change in an ecosystem. A direct driver unequivocally influences ecosystem processes. An indirect driver operates more diffusely, by altering one or more direct drivers.

Ecosystem A system of interacting living organisms together with their physical environment, which can range from a very small area to the entire earth.

EEZ (Exclusive Economic Zone) The portion of the seas and oceans extending from the seaward edge of the state's territorial sea up to 200 nautical miles from its coast, within which states have the right to explore and exploit natural resources as well as to exercise jurisdiction over marine science research and environmental protection.

El Niño-Southern Oscillation (ENSO) El Niño, in its original sense, is a warm water current which periodically flows along the coast of Ecuador and Peru, disrupting the local fishery. This oceanic event is associated with a fluctuation of the inter-tropical surface pressure pattern and circulation in the Indian and Pacific oceans, called the Southern Oscillation. This coupled atmosphere-ocean phenomenon is collectively known as El Niño-Southern Oscillation, or ENSO. During an El Niño event, the prevailing trade winds weaken and the equatorial counter-current strengthens, causing warm surface waters in the Indonesian area to flow eastward to overlie the cold waters of the Humboldt Current. This event has great impact on the wind, sea surface temperature and precipitation patterns in the tropical Pacific. It has climatic effects throughout the Pacific region and in many other parts of the world.

Endemic A species which is only found in a given region or location and nowhere else in the world.

Erosion The detachment and movement of soil particles by natural forces, primarily water and wind. More broadly, erosion is the process of wearing away rocks, geologic, and soil material via water, wind, or ice (e.g. glaciers).

Estuary A semi-enclosed coastal body of water that has a free connection with the open sea, and within which sea water is diluted by freshwater land drainage; or an inlet of the sea reaching into a river valley as far as the upper limit of the tidal rise.

Extreme weather event An event that is rare within its statistical reference distribution at a particular place. Definitions vary, but an extreme weather event would normally be as rare as or rarer than the 10th or 90th percentile. By definition, the characteristics of what is called extreme weather may vary from place to place. An extreme climate event is an average of a number of weather events over a certain period of time, an average which is itself extreme (e.g. rainfall over a season).

Eutrophication The over-enrichment of an aquatic environment with inorganic nutrients, especially nitrates and phosphates (e.g. sewage or fertilizer runoff). This may result in the stimulation of algal growth, potentially leading to the oxygen depletion of the water.

Extreme weather event An event that is rare within its statistical reference distribution at a particular place. Definitions vary, but an extreme weather event would normally be as rare as or rarer than the 10th or 90th percentile. By definition, the characteristics of what is called extreme weather may vary from place to place.

An extreme climate event is an average of a number of weather events over a certain period of time, an average which is itself extreme (e.g. rainfall over a season).

Fishing:

Artisanal or small-scale Traditional fisheries involving fishing households (as opposed to commercial companies), using a relatively small amount of capital and energy, relatively small fishing vessels (if any), making short fishing trips, close to shore, mainly for local consumption.

Demersal Fish, or the capture of fish living in close relation and dependence with the ocean bottom, such as cod, grouper and lobster. Much of such fishery is destructive. Deep-sea demersal trawling gear is large and heavy and has a very substantial impact on the seabed environment as a single deep-sea trawler may rake over 10 square kilometers (3 square miles) of seabed, daily. The general effects of trawling include extinction of species that form the habitat-structure, decrease in biodiversity, and a loss of long-lived organisms.

Pelagic Fish, and the capture of fish, that spend most of their life swimming in shoals in the water column with little contact or dependency on the bottom. Species range from the small, such as anchovies, sardines, and mackerel to the large, such as tuna.

Global warming Increase in global mean temperature.

Harmful Algal Blooms (HABs) Algal growth that is sufficiently dense to accumulate into visible patches near the surface of the water. Some species produce potent neurotoxins that can be distributed through the food chain.

Hypoxia Low in oxygen, the condition in which dissolved oxygen is below the level necessary to sustain most animal life.

Ice sheet A mass of ice that is sufficiently deep to cover most of the underlying bedrock topography. There are only two large ice sheets in the modern world, on Greenland and Antarctica.

Ice shelf A floating sheet of ice of considerable thickness attached to a coast (usually of great horizontal extent with a level or gently undulating surface); often a seaward extension of ice sheets.

Inorganic pollutants Inorganic chemical pollutants are naturally found in the environment but due to human development are often concentrated and released into the oceans in urban stormwater. The primary inorganic pollutants of concern are cadmium, copper, lead, zinc, nitrogen, nitrate, nitrite, ammonia, phosphorus, and phosphate.

Invasive species A species that does not naturally occur in a specific area and whose introduction causes, or is likely to cause, economic or environmental harm or harm to human health. Invasive species are seen as a prime threat to biodiversity by the IUCN.

Mariculture Cultivation, management, and harvesting of marine organisms in the sea in specially constructed rearing facilities e.g. cages, pens and long-lines. For the purpose of FAO statistics, mariculture refers to cultivation of the end product in seawater even though earlier stages in the life cycle of the concerned aquatic organisms may be cultured in brackish water or freshwater or captured from the wild.

Nutrients Pollution, especially by nitrogen and phosphorus, has consistently ranked as one of the top causes of degradation in some US waters for more than a decade. Excess nitrogen and phosphorus lead to significant water quality problems including harmful algal blooms, hypoxia, and declines in wildlife and wildlife habitats.

Persistent Organic Pollutants Chemical substances that persist in the environment, bioaccumulate through the food web, and pose a risk of causing adverse effects to human health and the environment.

Phytoplankton Or algae, are tiny plants that live in the upper layers of the ocean and form the basis of the marine food chain. In particularly nutrient-rich conditions (including eutrophication) phytoplankton blooms may occur and could be toxic.

Sea level rise Due to either thermal expansion or the melting of glaciers and land ice. Thermal expansion is the increase in volume (and decrease in density) that results from warming water. A warming of the ocean leads to an expansion of the ocean volume and hence an increase in sea level. The melting of ice sheets or ice already in the oceans will not add to sea levels, however, the melting of land ice results in runoff of melt-water into the seas that directly adds to the volume of the oceans and therefore sea levels.

Sedimentation The process in which particulate matter carried from its point of origin by either natural or human-enhanced processes is deposited elsewhere on land surfaces or in waterbodies. Sediment is a natural product of stream erosion, however, the sediment load may be increased by human practices. Such enhanced sources of sediment in a watershed include uncovered soil regions, such as construction sites, deforested areas, and croplands.

Storm surge The temporary increase, at a particular locality, in the height of the sea due to extreme meteorological conditions (low atmospheric pressure and/or strong winds). The storm surge is defined as being the excess above the level expected from the tidal variation alone at that time and place.

Thermohaline circulation Large-scale, density driven circulation in the ocean, caused by differences in temperature (thermo) and salinity (haline).

SUPPORTING MATERIAL FOR MARINE ECOSYSTEMS IN PERIL (pp. 34–35)

The material described was derived from the study reported in "A Global Map of Human Impact on Marine Ecosystems" by Benjamin S. Halpern, et. al. and published in *Science*. The survey examined the following drivers and ecosystems:

DRIVERS

Nutrients (fertilizer)
Organic pollutants (pesticides)
Inorganic pollutants
Direct human (population density)
Pelagic, low-bycatch fishing
Pelagic, high-bycatch fishing
Demersal, destructive fishing
Demersal, non-destructive, low-bycatch fishing
Demersal, non-destructive, high-bycatch fishing
Artisanal fishing
Oil rigs
Invasive species
Ocean pollution
Shipping
Sea surface temperature (changes due to climate change)
Ultraviolet radiation (reaching earth)
Ocean acidification

ECOSYSTEMS

Coral see pp. 48–49

Seagrass see pp. 44–45

Mangrove see pp. 46–47

Rocky reef Rocks lying at or near the sea surface.

Shallow soft Areas up to 60 meters in depth with a soft sediment floor.

Hard shelf A gently sloping area extending from the low-water line to the depth of a marked increase in slope around the margin of a continent or island, 60–200 meters deep with a hard substrate.

Soft shelf As for a hard shelf, but with a soft sediment floor.

Hard slope A relatively steeply sloping surface lying seaward of the continental shelf, 200–2,000 meters deep with a hard substrate.

Soft slope As for a hard slope, but with a soft sediment floor.

Hard deep Waters deeper than 2,000 meters with a hard substrate.

Soft deep As for hard deeps, with a soft sediment floor.

Seamounts Undersea mountains whose summits lie beneath the ocean waves. They are usually volcanic in origin and are generally defined as having an elevation of greater than 1,000 meters from the seabed.

Pelagic waters The open water environment, up to 60 meters depth in all areas where the ocean is deeper than 60 meters.

Deep waters The water column below 60 meters in depth.

Singapore's industrial harbor: Population and commercial pressures can lead to the wholesale loss of coastal ecosystems.

Part One: People and Coasts

Population Growth along Coasts

About half of the world's population, some 3.2 billion people, live along or within 200 kilometers (124 miles) of a coastline, on just 10 percent of the earth's land area.

People have lived along coastlines and inland seas since the dawn of time. As natural centers of commerce and trade, coastal towns quickly evolved into cities. This trend has accelerated in the past half century, as rural economies atrophy and more people migrate to towns and cities in search of employment and access to basic services. This is having profound consequences for coastal and near shore environments.

Population densities along coastlines now average 80 people per square kilometer (207 per square mile), but in many of the world's crowded urban corridors along coastlines, such as the Nile Delta in Egypt, and the sprawling urban agglomerations of Mumbai and Kolkata in India, population densities exceed 1,000 people per square kilometer (2,588 per square mile). The figures for Asia are staggering. Nearly 1,000 people arrive every day in China's large coastal cities, such as Shanghai, Tianjin, and Shenzhen. Similar numbers are heading to urban coastal areas in Vietnam, the Philippines, Cambodia, Thailand, and Bangladesh.

Global tourism has fuelled booming local economies, but in some cases has proved environmentally unsustainable, as infrastructure is built too

POPULATION OF US COASTAL COUNTIES

estimated

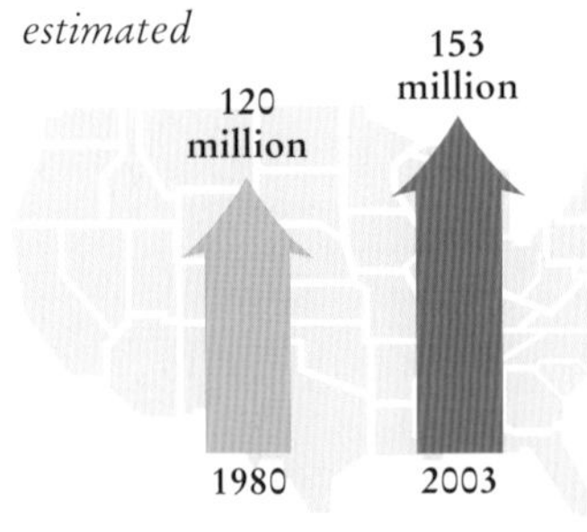

Manila is dominated by slums, such as this one, not far from the business district. The city's canals are nothing more than open sewers which transport enormous quantities of untreated wastes into coastal waters.

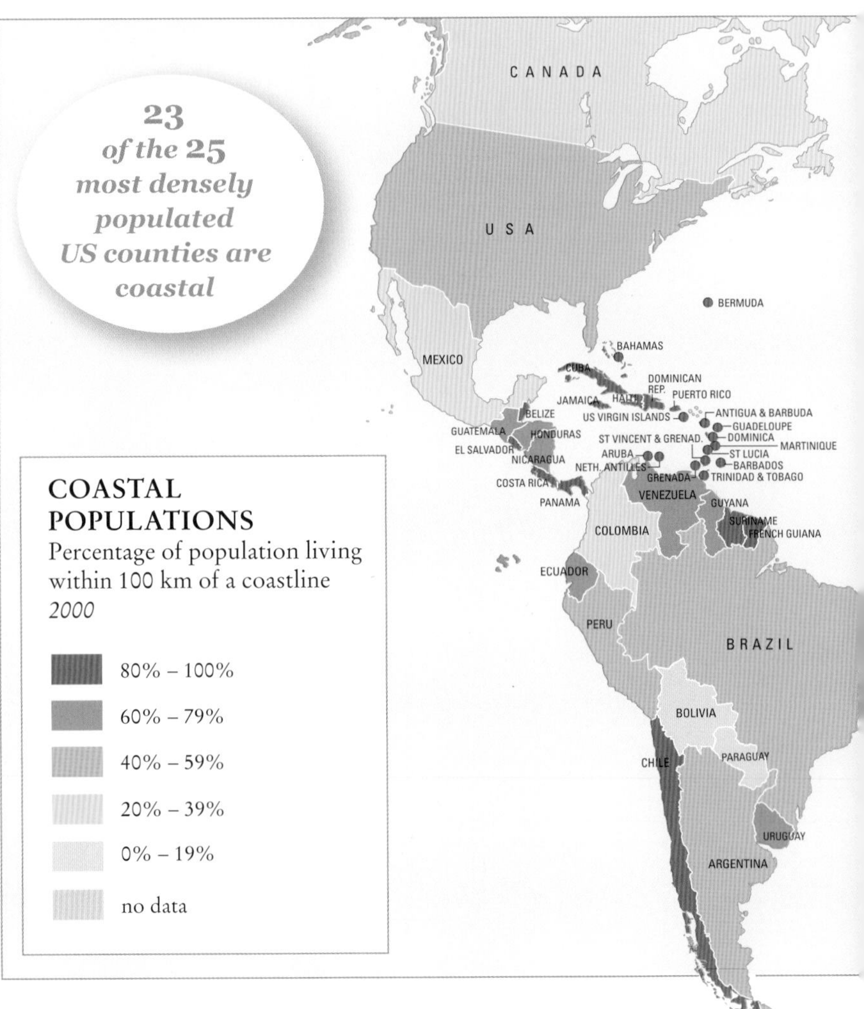

close to fragile coastal ecosystems such as sandy beaches, wetlands, seagrass beds, and coral reefs. The problems of high coastal population density in the Mediterranean, for instance, are compounded by the popularity of the region as a holiday destination, resulting in additional seasonal pressure on water resources and the natural environment, as well as generating enormous quantities of waste.

Globally, the wastes of coastal and near shore populations continue to have a serious impact on coastal resources. Scientists have estimated that 80 percent of all marine pollution comes from land-based sources.

Toxic chemicals and heavy metals that leach into coastal waters from industrial and agricultural activities can damage human health and poison marine life. Persistent organic pollutants (POPs), in particular, tend to bio-accumulate up the food chain over time and can result in reproductive, immunological, and neurological problems in both marine life and humans. Beluga whales found in the mouth of Canada's St. Lawrence Seaway have such high levels of PCBs in their blubber that under Canadian law they qualify as "toxic waste dumps".

POLLUTED OCEANS

Proportion of untreated wastewater discharged into coastal waters in developing countries
2010

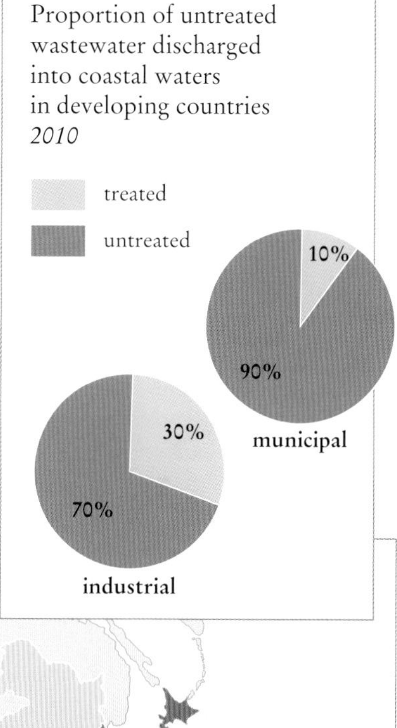

Urbanized Coastlines

For the first time in history the majority of the world's people live and work in towns and cities.

In 1800 only 2 percent of the world's population was urban. By 2008 this had increased to 50 percent (3.3 billion people). The United Nations projects that by 2030, 60 percent of the global population will be urbanized. The urban populations of Africa and Asia will grow most quickly, doubling between 2000 and 2030. Most growth will be due to two factors: the influx of mostly poor people from rural areas, and the expected higher fertility levels within urban slums and squatter settlements.

It is the rapidity and unplanned nature of this anticipated urban explosion that worries demographers and urban planners. Most of this growth in developing countries will be chaotic and unplanned, and will increase pressure on already woefully inadequate infrastructures. Urban water supplies in developing countries are often seriously contaminated by discharges of untreated sewage and industrial wastes, and poor solid waste management. Worldwide, two-thirds of the sewage from urban areas is pumped untreated into lakes, rivers, and coastal waters, with detrimental consequences for health and the environment.

GROWING URBANIZATION

Percentage of population living in urban centers
1950–2050

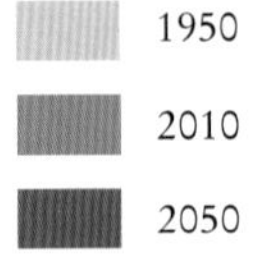

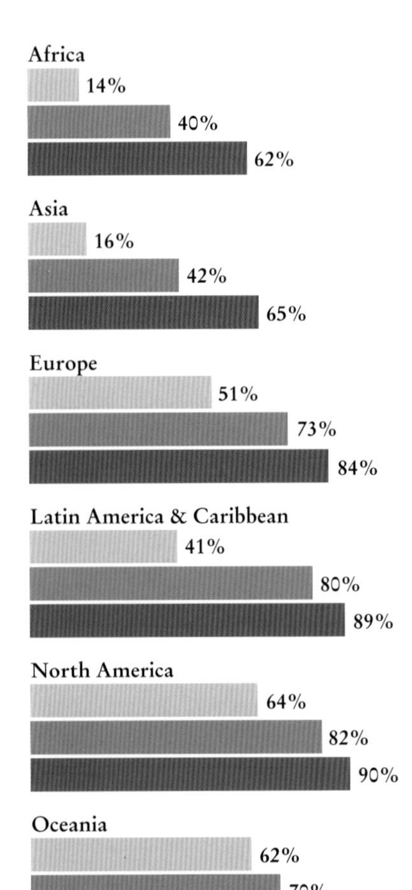

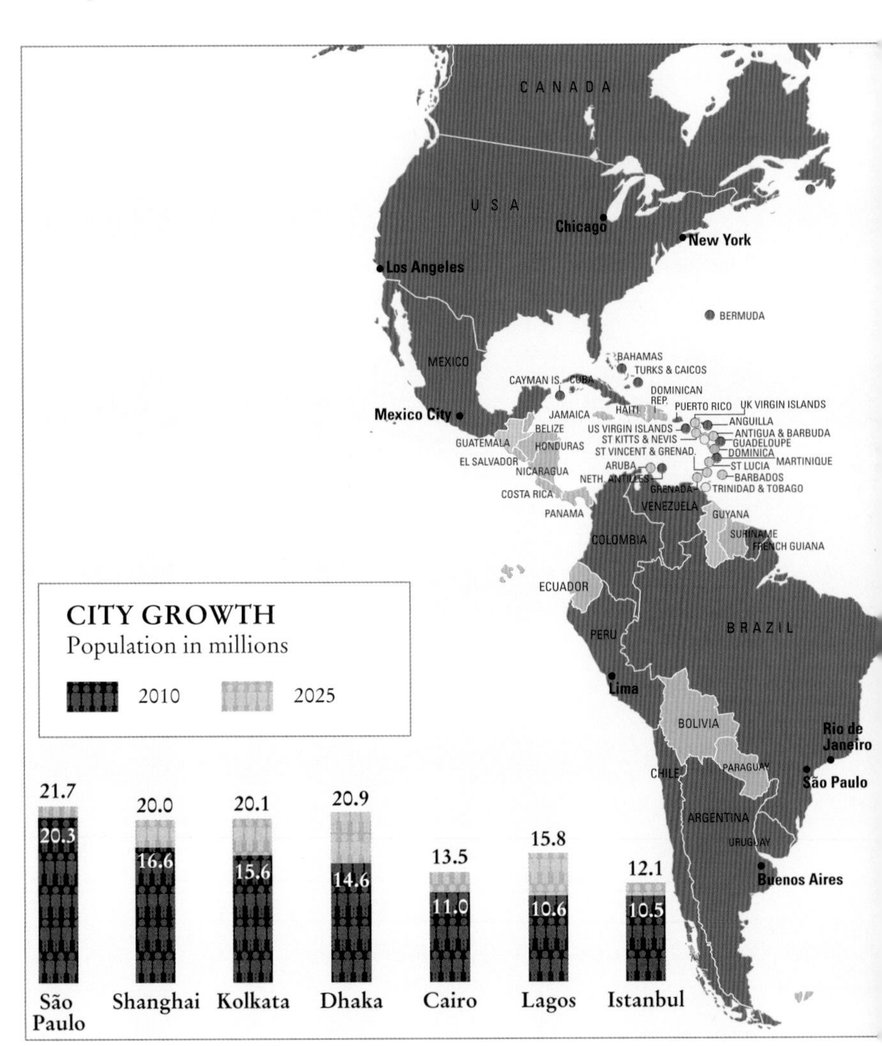

 ◀◀ 24–25

Urban energy consumption also creates heat islands that alter local weather patterns and increase rainfall. However, the built environment reduces water infiltration and lowers water tables, while increasing runoff, adding to flooding and pollution problems. Flooding will become more severe as sea levels rise, and extreme weather events may occur more frequently. Populations concentrated along coastlines will be very vulnerable to these changes.

Since coastal cities are growing rapidly in most developing countries, urban sprawl will have an enormous impact on vulnerable ecosystems. Increasing urbanization of the west coast of Trinidad has resulted in much higher levels of marine pollution and alteration of near shore currents, leading to changes in sediment deposition and increases in both coastal erosion and accretion. In Europe loss of coastal wetlands such as salt marshes, seagrass beds, and shellfish reefs is resulting in declining food resources and extinction of species, as well as increasing coastal vulnerability to storm surges and erosion.

URBAN WORLD

Percentage of population living in urban centers

2010

- 75% – 100%
- 50% – 74%
- 25% – 49%
- 0% – 24%
- no data
- • thirty largest cities *2010*

28–29, 30–31, 62–63, 64–65, 74–75

Urbanized Coastlines: Sub-Saharan Africa

Sub-Saharan Africa has the fastest-growing population of any region in the world. By 2050 it is projected to have twice as many people as in 2009 – nearly 1.7 billion – with an increasing number moving to capitals and coastal cities.

Population growth and skewed distribution patterns are already straining coastal and near-shore environments throughout the continent, resulting in the degradation of valuable coastal ecosystems. Mangroves are being cleared for fuel, charcoal, and building materials; coral reefs are being mined for limestone; and seagrasses and coral reefs smothered by millions of tons of soil that has been washed off deforested land by rain, and carried to the coast by river systems. In Madagascar, 90 percent of the island's cooking fuel is wood and agricultural waste, and mangrove forests have been decimated by fuelwood collectors who sell the wood for fuel or convert it into charcoal. In West Africa, where mangrove forests have been reduced to a quarter of their original size, mangrove wood is boiled to extract salt from it.

Dar es Salaam is growing by 4.3% a year which will double its population of 3.3 million in one generation

Loss of productive coastal ecosystems and the biodiversity they harbor, coupled with rampant over-fishing has led to a decline in the harvest of wild fish. This has meant that people have lost their livelihood and an important source of food. Over the past two decades, fish consumption per capita in much of Sub-Saharan Africa has declined, while efforts to harvest fish have intensified. This is due mainly to a failure by governments to instigate coastal management programmes, or to enforce fishing regulations.

Although West African fisheries, for instance, are worth some $3 billion a year, most of the value is taken by the fishing fleets of foreign nations, as well as by illegal fishing operations. The fees paid by foreign fleets, mostly from the EU and Russia, to the governments of the countries concerned, are far below fair value, amounting to no more than $200 million a year, robbing coastal states of both income and a badly needed source of protein.

Pirate fishing operations, in particular, are increasing pressure on fish stocks. A study carried out in 2006 in Guinea tracked 92 trawlers: a third of them had no licenses to fish in Guinea's territorial waters; one in 10 had no name or flag; and half were fishing illegally at least some of the time (taking more fish than allowed, fishing in protected areas, or going after species not included in their agreements).

URBAN POPULATION GROWTH
2005–2010

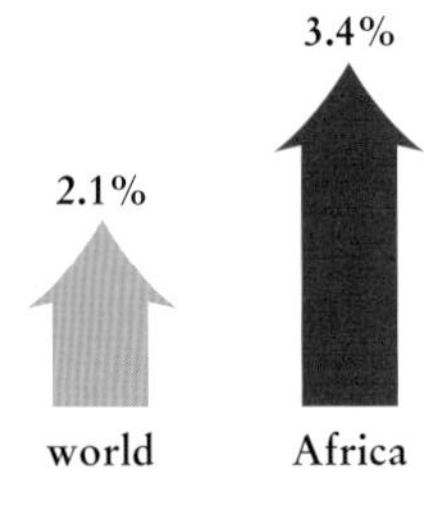

GROWING COASTAL CITIES
2007 & population projection 2025
millions

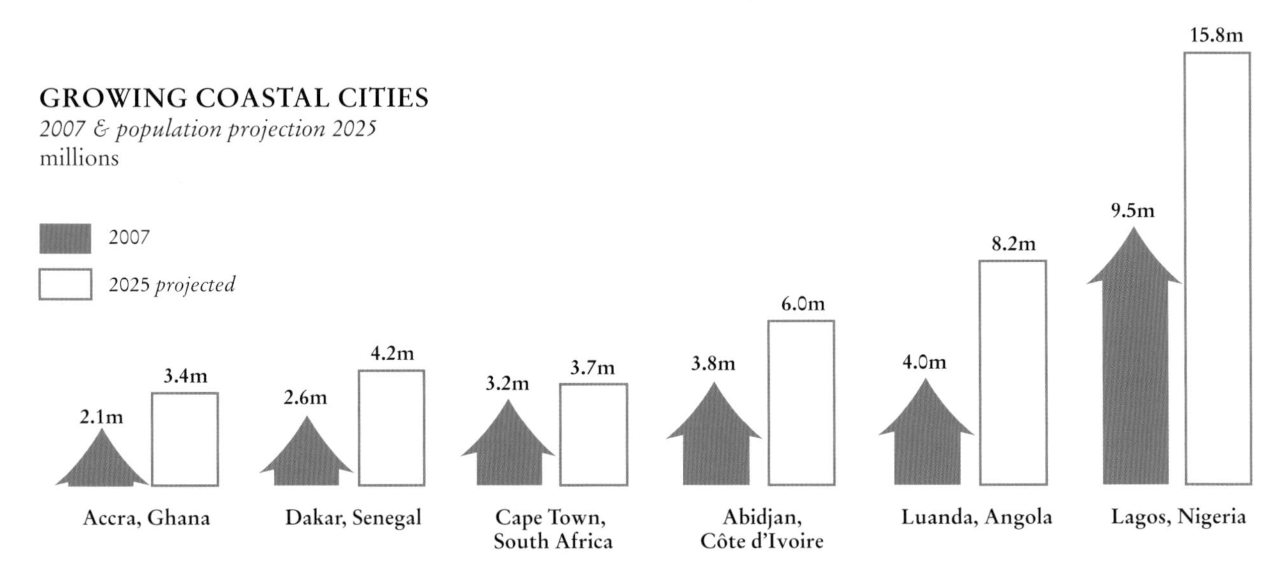

24–25, 26–27

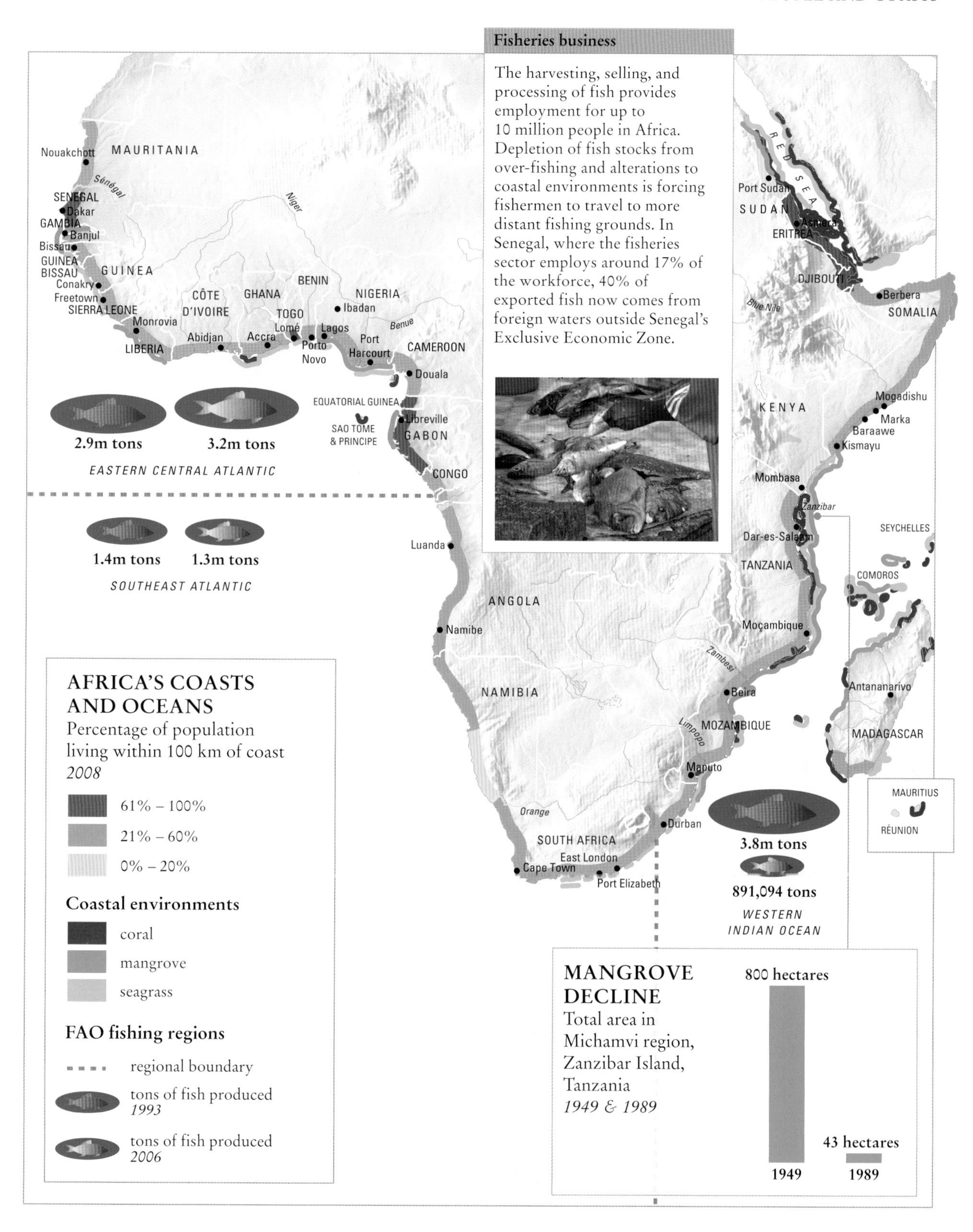
Fisheries business
The harvesting, selling, and processing of fish provides employment for up to 10 million people in Africa. Depletion of fish stocks from over-fishing and alterations to coastal environments is forcing fishermen to travel to more distant fishing grounds. In Senegal, where the fisheries sector employs around 17% of the workforce, 40% of exported fish now comes from foreign waters outside Senegal's Exclusive Economic Zone.
AFRICA'S COASTS AND OCEANS
Percentage of population living within 100 km of coast
2008
61% – 100%
21% – 60%
0% – 20%
Coastal environments
coral
mangrove
seagrass
FAO fishing regions
regional boundary
tons of fish produced 1993
tons of fish produced 2006
2.9m tons
3.2m tons
EASTERN CENTRAL ATLANTIC
1.4m tons
1.3m tons
SOUTHEAST ATLANTIC
3.8m tons
891,094 tons
WESTERN INDIAN OCEAN
MANGROVE DECLINE
Total area in Michamvi region, Zanzibar Island, Tanzania
1949 & 1989
800 hectares
43 hectares
1949
1989
MAURITANIA
Nouakchott
Sénégal
SENEGAL
Dakar
GAMBIA
Banjul
Bissau
GUINEA BISSAU
GUINEA
Conakry
Freetown
SIERRA LEONE
Monrovia
LIBERIA
CÔTE D'IVOIRE
Abidjan
GHANA
Accra
TOGO
Lomé
BENIN
Porto Novo
Lagos
Ibadan
NIGERIA
Niger
Benue
Port Harcourt
CAMEROON
Douala
EQUATORIAL GUINEA
SAO TOME & PRINCIPE
Libreville
GABON
CONGO
Luanda
ANGOLA
Namibe
NAMIBIA
Orange
SOUTH AFRICA
Cape Town
East London
Port Elizabeth
Durban
Maputo
Limpopo
MOZAMBIQUE
Beira
Zambezi
Moçambique
TANZANIA
Dar-es-Salaam
Zanzibar
Mombasa
KENYA
Kismayu
Baraawe
Marka
Mogadishu
SOMALIA
Berbera
DJIBOUTI
ERITREA
Asmara
SUDAN
Port Sudan
RED SEA
Blue Nile
SEYCHELLES
COMOROS
Antananarivo
MADAGASCAR
MAURITIUS
RÉUNION

Eroding Shorelines

Coastal erosion occurs when winds, waves, and currents scour sand and soil from coastal areas, depositing it offshore. This is a natural phenomenon but is aggravated by human activities.

Coastal erosion is a result of the mercurial nature of coastlines constantly under assault by the forces of sea and weather. In the last fifty years, several factors have accelerated coastal erosion processes including the uncontrolled expansion of coastal cities and towns; the filling in and draining of wetlands; the damming of rivers and streams; the mining of sand and coral for building materials; and sea level rising triggered by global climate change.

By far the worst effects have been due to the construction of dams and river diversions in watersheds, robbing coastlines of replenishing soil. Over the course of the twentieth century, some 70 percent of the world's sandy beaches were in retreat. One of the worst examples of coastal land loss is the US state of Louisiana. As a result of the damming of the Mississippi River and its tributaries the amount of sediment reaching the Delta region has plunged from 400 million tons in 1900 to around 140 million tons today.

The rest of the world's coastlines have fared little better. Overall, Europe's coastlines are in retreat: 20 percent are eroding or in need of protection measures. Serious coastal erosion has afflicted many Asian countries. Sri Lanka's coastal land loss has been attributed to the wholesale destruction of mangrove swamps, which were whittled down from 12,000 hectares in 1986 to under 6,000 hectares by 2000. In Australia, Gold Coast beaches are eroding rapidly due to sea level rises and an increase in tropical storms, phenomena made worse by global climate change.

STATE OF EUROPE'S COASTS

Total length of coast 100,925 km (62,711 miles)

artificially stabilized 5%
protected but eroding 3%
unprotected & eroding 12%
accreting 14%
naturally stable 39%
no data or not applicable (harbors, estuaries) 27%

15 km²
of coastal land is lost to the sea every year in the European Union

EUROPE'S ERODING COASTLINE

Percentage of coastal land eroding
2004

30% – 55%
20% – 29%
10% – 19%
0% – 9%
no data

◀◀ *24–25, 26–27*

Developed countries have constructed expensive groynes, bulkheads, and retaining walls, and initiated beach nourishment projects, in an attempt to minimize erosion. Beach nourishment is a Sisyphean exercise – a continuing expense which addresses the consequences instead of the causes. The city of Miami Beach, Florida, for instance, spent $64 million restoring the city's famed beaches between 1976 and 1981. However, with the beaches continuing to erode at the rate of 4 to 7 meters a year, the city is now sinking artificial reefs offshore to reduce wave action and slow down erosion rates.

In the USA, California's Big Sur coastline is vulnerable to erosion and landslides due to a combination of recurrent uplift caused by earthquakes creating faults and fractures, heavy winter rainfall, high wave energy, and loss of vegetation as a result of summer wildfires.

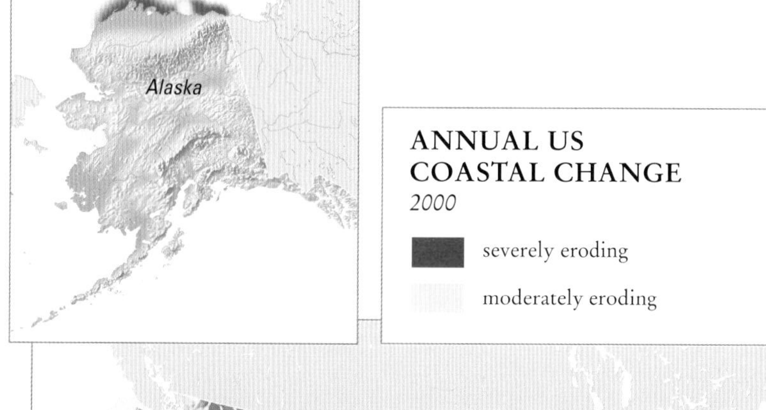

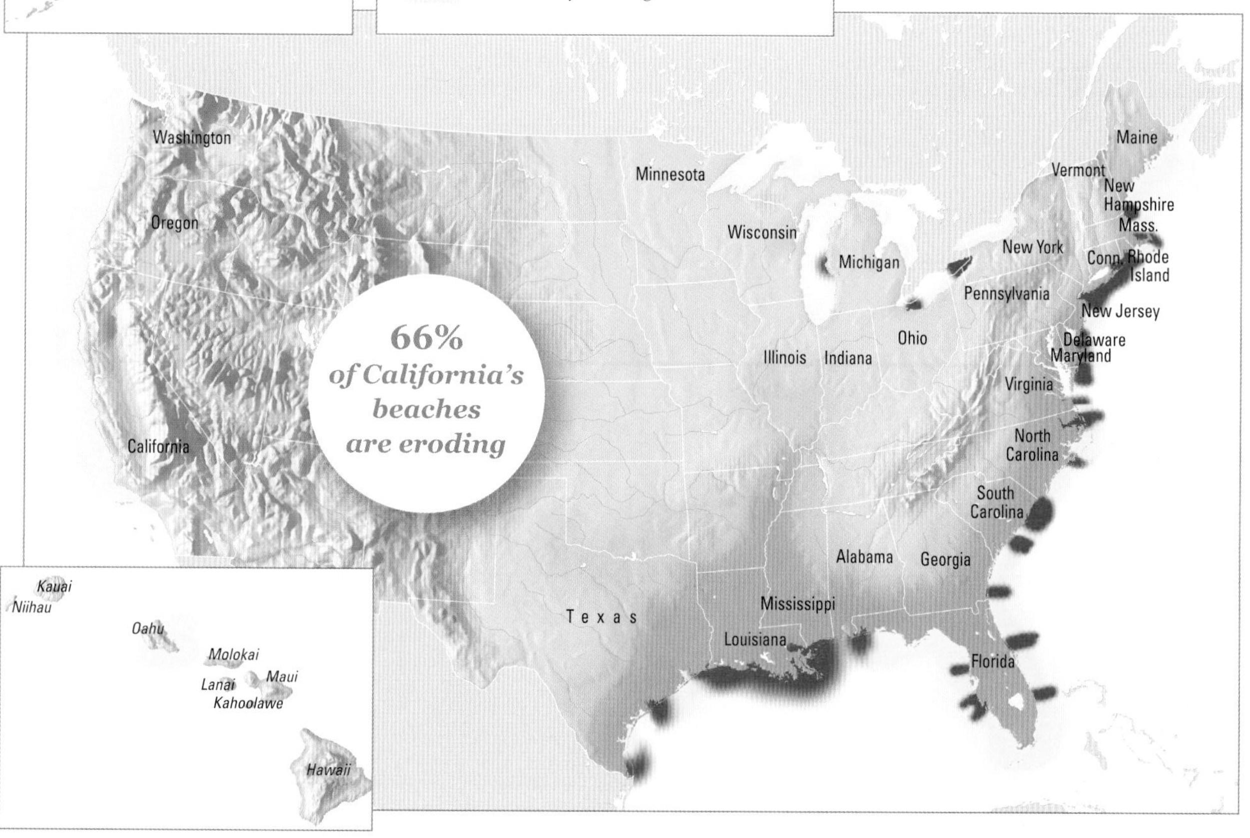

Tidal saltmarsh in Northern Germany: Saltmarshes are one of the richest, yet most endangered, coastal ecosystems.

Part Two: Major Threats to Ocean Resources

Marine Ecosystems under Threat

Human impacts over the past 60 years have led to a precipitous decline in the viability of coastal and ocean ecosystems; vital functions on which the rest of the planet depends.

Sea nettle jellyfish have swarmed in huge numbers in recent years.

Scientists from a cross-section of disciplines have recently mapped the cumulative impact of human activity on the world's oceans. They examined anthropogenic drivers of ecosystem damage – including overfishing, fertilizer runoff, pollution from oil and gas extraction, untreated sewage, industrial wastes, shipping, and climate change – across 20 vital marine ecosystems, such as coral reefs, seagrass beds, mangrove swamps, continental shelves, and seamounts.

The results are alarming. Some 41 percent of the ocean is suffering a medium high to high impact from humanity's damaging activities (amounting to nearly 170 million square kilometres or 66 million square miles of ocean area). No single spot in the ocean remains unaffected by at least one of the 17 factors. The worst-hit areas include: the North Sea, South and East China Seas, the East Coast of North America, the Caribbean, Mediterranean and Red Seas, the Bering Sea and parts of the western Pacific (around Japan), and the Persian Gulf.

Marine ecosystems most under threat include: coral reefs, nearly half of which have suffered severe impacts; seagrass beds; mangrove swamps in estuaries; rocky reefs; and shallow continental shelves near population centres, such as those found around the North and Mediterranean Seas.

Only four percent of the ocean remains relatively unaffected. Ocean depths, shallow areas with muddy bottoms, and remote coral reefs are apparently least affected by human activities.

The polar seas are among those areas experiencing only low impacts from most of the drivers examined. However, due to the fact that cold waters absorb more CO_2 than warmer ones, they are disproportionately affected by rising levels of CO_2 in the atmosphere, resulting in acidification. It is also feared that as the climate continues to change, the areas of sea ice will dwindle, opening up more of the polar seas to destructive fishing, as well as oil and gas exploitation.

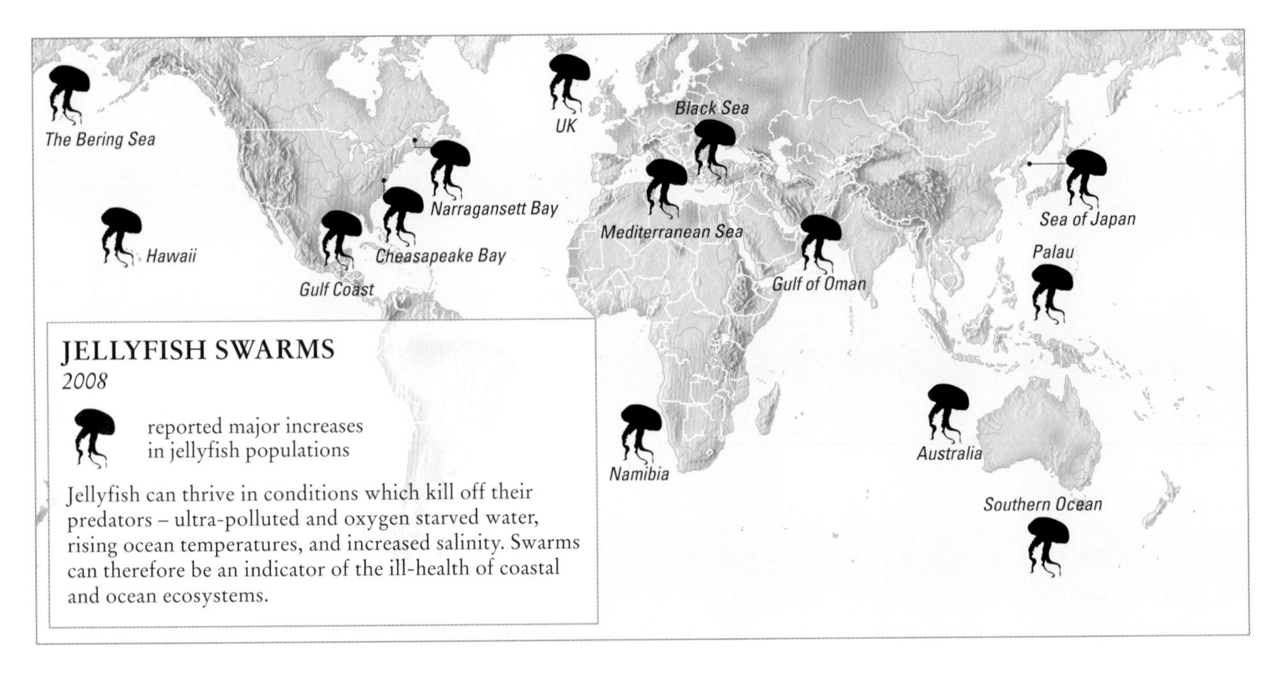

◀◀ 21, 24–25

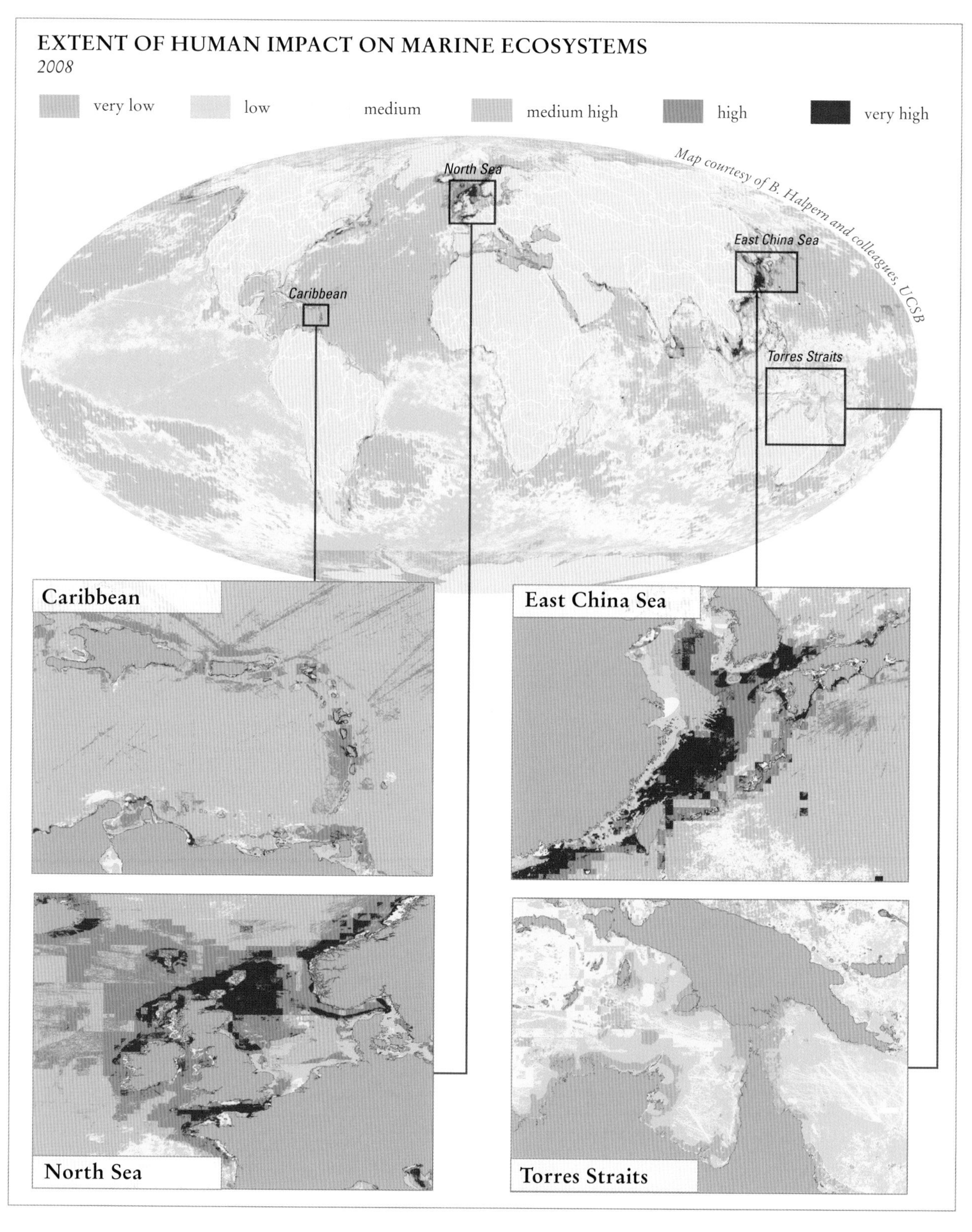
EXTENT OF HUMAN IMPACT ON MARINE ECOSYSTEMS
2008
very low
low
medium
medium high
high
very high
Map courtesy of B. Halpern and colleagues, UCSB
North Sea
East China Sea
Caribbean
Torres Straits
Caribbean
East China Sea
North Sea
Torres Straits

Ocean Dead Zones

The number of coastal dead zones has doubled every decade since 1960.

Coastal zones around the world's oceans and seas are increasingly on the receiving end of a glut of nutrients which have been washed off the land. These consist mainly of nitrogen and phosphorus from agricultural fertilizers, untreated sewage, and organic industrial wastes, and they turn near-shore ecosystems into marine graveyards, where little life, other than bacteria and other microbes, survives.

The nutrients stimulate the runaway growth of algae and plankton on the surface. When these organisms die and sink to the bottom, they are devoured by microbes, which consume oxygen in the process. The result of this process of eutrophication is an hypoxic area of "dead sea"; while fish and crustaceans can usually swim out of dead zones, slow-moving or immobile sea life, including sea anemones, sponges, and tube-worms, as well as bivalve molluscs such as oysters and clams, are unable to relocate so easily and often perish.

LOSS OF POTENTIAL FOOD FOR FISH

Estimated annual weight of organisms lost

2008

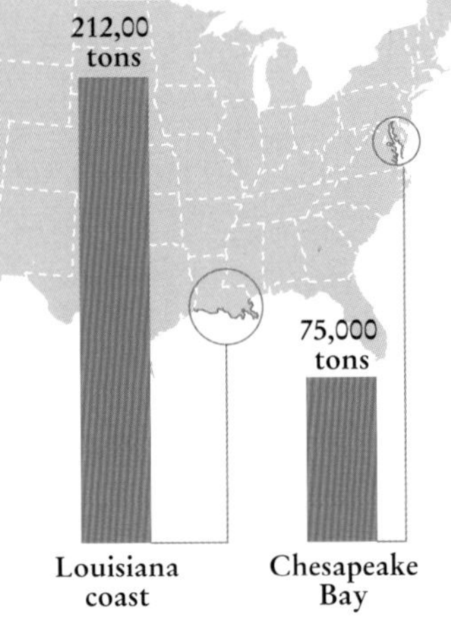

HYPOXIA AND EUTROPHICATION

2010

hypoxic area 10,000 km² or larger

2008

hypoxic area

evidence of eutrophication

area showing recovery from hypoxia

Gulf of Mexico

In 2010, the dead zone in the Gulf of Mexico, off the coast of Louisiana and Texas reached its fifth largest recorded extent – 19,999 square kilometers (7,722 square miles). The dead zone is caused by the continuous glut of nutrients and other pollutants deposited by the Mississippi River's massive watershed, which drains nearly two-thirds of the continental USA. The red and orange areas represent high concentrations of phytoplankton and river sediment which can lead to hypoxia (loss of oxygen).

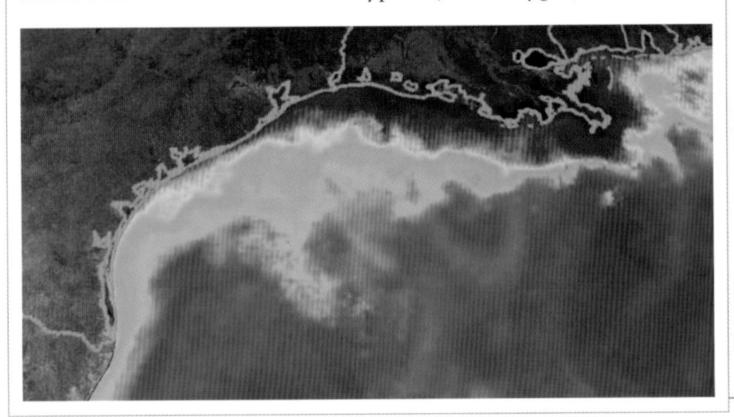

Over 400 such dead zones have been identified around the world's coastlines. Most of these dead zones are found in the waters of developed countries, and many of them in prime fishing grounds. The overwhelming majority are the result of land based pollution runoff. However, in recent years researchers have uncovered evidence that climate change is contributing to the spread of some dead zones, such as the one off the Oregon coast in the USA.

The lack of food sources in dead zones leads to a decline in fish populations and consequently, reduced catches. The resulting losses to the fishing industry total millions of dollars a year. Chesapeake Bay and the Gulf of Mexico – where the Mississippi deposits a tremendous load of nitrogen-rich fertilizers and other nutrients scoured off the farmlands in its huge watershed – are North American coastal areas that are particularly badly affected.

Climate change and dead zones

Research has linked climate change and rising ocean temperatures to an increase in dead zones. A new computer model developed by Danish researchers indicates that climate change could make dead zones a permanent fixture of coastal waters. Global climate change and rising ocean temperatures reduce the ocean's ability to store oxygen, while at the same time decreasing the amount of oxygen available in deep water. Though current dead zones make up no more than two percent of the world's ocean volume, they could grow by a factor of ten or more by 2100.

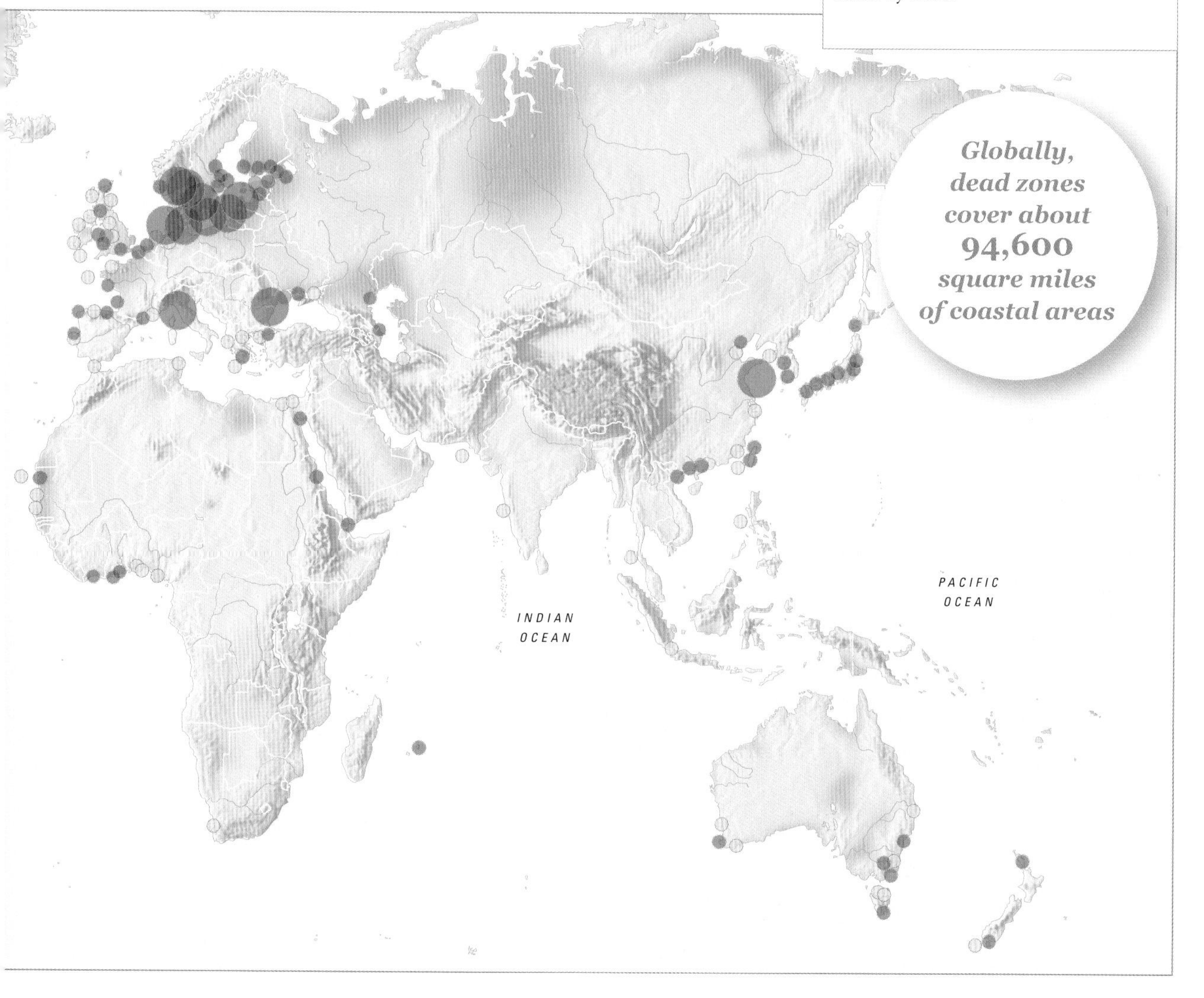

38–39, 40–41

Ocean Dead Zones: The Baltic Sea

The almost-enclosed waters of the Baltic are particularly vulnerable to pollution.

The Baltic is a sea of superlatives. It is the youngest sea, having reached its present form only 3,000 years ago. It is the shallowest sea, averaging just 53 meters in depth. And, apart from the Black Sea, it is also the most enclosed sea. It takes 30 years for its waters to be renewed through the narrow Danish Straits, its only link to the North Sea and the world ocean, while a vast river system contributes 2 percent of its volume as fresh water each year. This makes it the largest body of brackish water on the planet.

The Baltic's catchment area is home to at least 85 million people, 15 million of whom live within 10 kilometers (6 miles) of the coast. In the more industrialized, heavily urbanized areas of Poland, Germany, and Denmark, the population density is nearly 500 people per square kilometers (1,400 per square mile).

The very characteristics – confined, shallow, and brackish – that make the Baltic unique are contributing to its demise. Nutrient pollution, mostly from nitrogen and phosphorous, is degrading and impoverishing the sea's ecosystems. Most of the pollution originates from untreated or partially treated sewage, municipal wastes, and agricultural runoff, predominately from fertilizers and animal wastes. Since 1900, nitrogen levels in the Baltic have increased four-fold, and phosphorus levels eight-fold, in line with intensified agricultural activities.

The Baltic is also on the receiving end of thousands of tons of toxic pollutants, such as DDT and PCBs (known as persistent organic pollutants or POPs), as well as heavy metals from mining and industry, toxic chemicals from the pulp and paper industry, hydrocarbons from gas and oil and wastes from hospitals, tanneries, metal smelters and other sources.

DEAD AND TOXIC

Highest contributors of waterborne lead to Baltic Sea
averages 2000–01 & 2005–06
tons per year

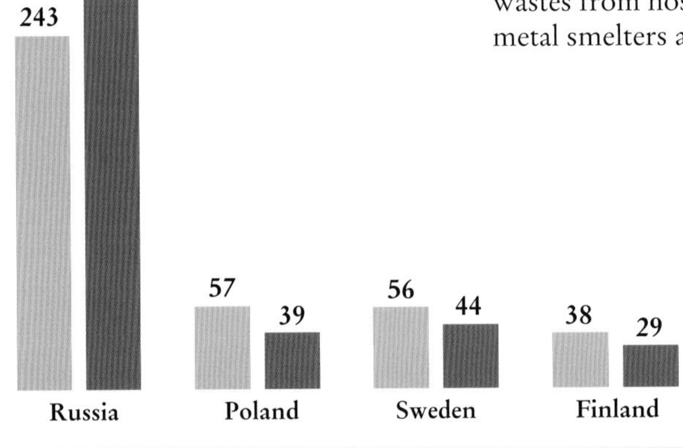

HELSINKI CONVENTION

Efforts to protect the Baltic date from 1974, when the states surrounding the sea (which included the former Soviet Union and the GDR) gathered to sign the Helsinki Convention on the Protection of the Marine Environment. It also created the Baltic Marine Environment Protection Commission, known as the Helsinki Commission (HELCOM) to oversee its implementation.

Progress has been slow but steady. Of the 149 regional pollution hot spots identified in 1990, 75 remained in 2009. Concentrations of DDT and PCBs have fallen, and anticipated investments in clean-up operations and pollution control are expected to eliminate half a million tons of biological oxygen demand (BOD) by 2012.

However, the build-up of phosphorus and nitrogen in the Baltic continues, albeit at a slower rate, transforming its deep waters into an anoxic desert with no life-form higher than bacteria in water below 165 feet (50 m).

NITROGEN AND PHOSPHORUS

Total waterborne loads
1990 & 2006
tons

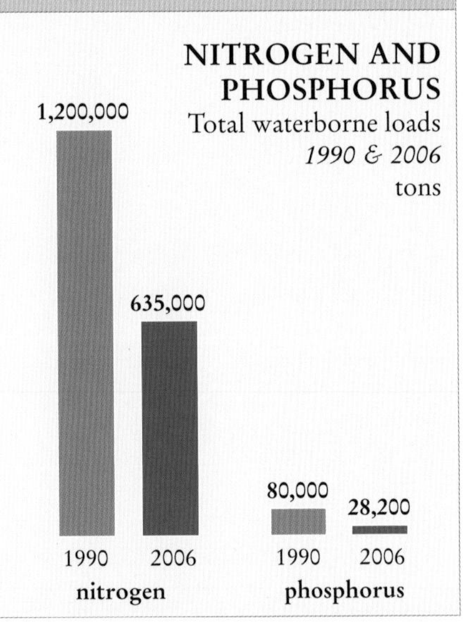

36–37

Fisheries

Half of Baltic's commercially important fish species are at risk, with populations below the critical biological level for survival. Cod stocks have never recovered their pre-1980 levels and are in permanent decline. In all some 85 species, including salmon, Baltic eels, shad, cod, haddock, pollack, mackerel, and tuna are threatened. Salmon and herring have recovered a little in recent years, but are still off limits to pregnant women because of high levels of dioxins in their flesh.

PRESSURES ON THE BALTIC

2009

● cities of over 500,000 people

● largest eutrophication hotspots *2009*

NORWEGIAN SEA
NORWAY
SWEDEN
FINLAND
Gulf of Bothnia
Helsinki
St. Petersburg
Gulf of Finland
Oslo
Stockholm
ESTONIA
RUSSIA
Skagerrak
Göteborg
Gulf of Riga
Riga
LATVIA
Kattegat
DENMARK
Danish Straits
Copenhagen
BALTIC SEA
LITHUANIA
NORTH SEA
RUSSIA
Vilnius
GERMANY
Hamburg
BELARUS
POLAND

Effect of POPs on bird health

There is a strong relationship between the reproductive ability of the fish-eating white-tailed sea eagle and concentrations of DDTs and PCBs in their eggs. In the 1970s, these majestic Baltic birds had only one-fifth the number of chicks as in pre-1950. Productivity began to improve in the 1980s, following bans on DDT and PCB use, and by the mid-1990s had largely recovered. The population of sea eagles on the Swedish Baltic coast has increased 8% per year since 1990.

Algal blooms

Since the 1970s, the Baltic has been plagued with seasonal algal blooms, but that of summer 2005 was one of the most severe and destructive. The algae were so thick that locals likened the effect to "rhubarb soup".

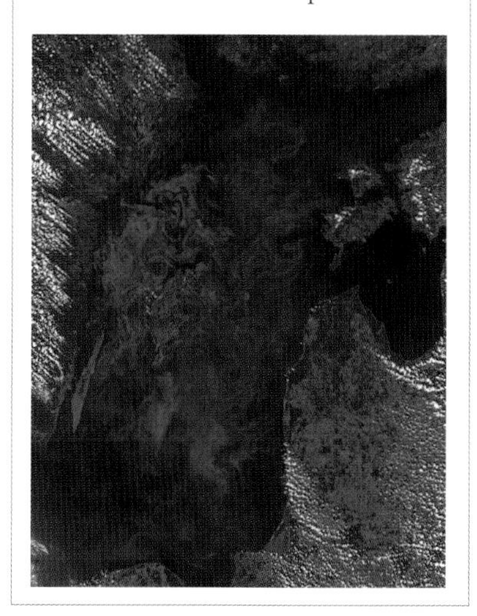

Mercury

In 1999, Poland was responsible for 104 tons of waterborne mercury flowing into the Baltic. In 2006, this had been reduced to 10 tons.

40–41 ▶▶

Ocean Dead Zones: The Northwest Pacific

The Northwest Pacific region encompasses the East China Sea (including the Bohai and Yellow Seas), the Sea of Japan, and the Okhotsk Sea in Pacific Russia. The region suffers from increasing red tides or harmful algal blooms (HABs) and the proliferation of dead zones. Scientists have identified and documented 24 dead zones, mostly along the coasts of China and Japan. Another nine areas, though not hypoxic yet, are of concern.

80% *of marine pollution in the region is from land based sources*

Seasonal HABs are an indicator of the ill health of coastal waters and encompass two phenomena: the water discoloration of red tides caused by huge numbers of unicellular phytoplankton that lead to deterioration of aquatic ecosystems and occasional fishery damage, and the proliferation of toxin-producing phytoplankton species that contaminate fish and shellfish throughout the food chain, and cause food-poisoning in people. Enhanced plankton production occurs when there is an ample supply of nutrients and so red tides are connected with the nutrient enrichment of waters (eutrophication). These HABs last for between seven days and 2 months. Damage is variable, but HABs can result in considerable environmental and socio-economic costs, including significant losses to the fishing industry.

The Sea of Japan appears to be particularly plagued with a disproportionate number of hypoxic areas, but this is because the Japanese have documented them more thoroughly than has China. Japan's coastal waters, including the inner shelf of the Yangtze River, are increasingly eutrophied as a result of the increasing influx of agricultural chemicals, and pollution from mining and mariculture operations and industry.

Along the south coast of the Korean Peninsula, the Nakdong Estuary and Masan Bay are highly contaminated with heavy metals and organic pollutants, such as PCBs and DDT, mostly from assorted industrial operations and agricultural runoff, both areas also suffer from HABs.

Further north, along the Russian coast, the Tuman River Delta and near shore areas are heavily contaminated with suspended solids and heavy metals pulled out of its huge watershed.

MAJOR ACTIVITIES AFFECTING COASTAL ZONES IN THE NORTHWEST PACIFIC

Activity	Major impact	Present status	Trend	Seas affected
Urbanization	Eutrophication	Major	Increasing	East China Sea (includes Bohai and Yellow Seas)
Agriculture	Eutrophication, pesticide pollution	Major	Increasing	East China Sea, Sea of Japan
Damming or diversion of rivers	Nutrient and sediment build up, coastal erosion	Major	Increasing	East China Sea
Deforestation	Erosion, sedimentation of coastal waters, habitat loss	Major	Decreasing	East China Sea
Aquaculture, mariculture	Eutrophication	Major	Increasing	East China Sea
Industrial growth	Pollution and HABs	Medium (major locally)	Increasing	East China Sea, Sea of Japan
Mining, including offshore oil & gas	Loss of habitat and biodiversity	Medium	Increasing	Sea of Japan

◀◀ 36–37, 38–39

Mapping Hypoxia

Chinese waters in the East China, Yellow and Bohai Seas are severely contaminated with untreated sewage and municipal wastes, as well as toxic industrial wastes and runoff from increasing loads of fertilizers, however, the data points are too coarse to identify specific regions. Instead the eutrophic and hypoxic areas of the seas are represented as one dot each, even though vast near shore stretches of both seas are considered biologically dead.

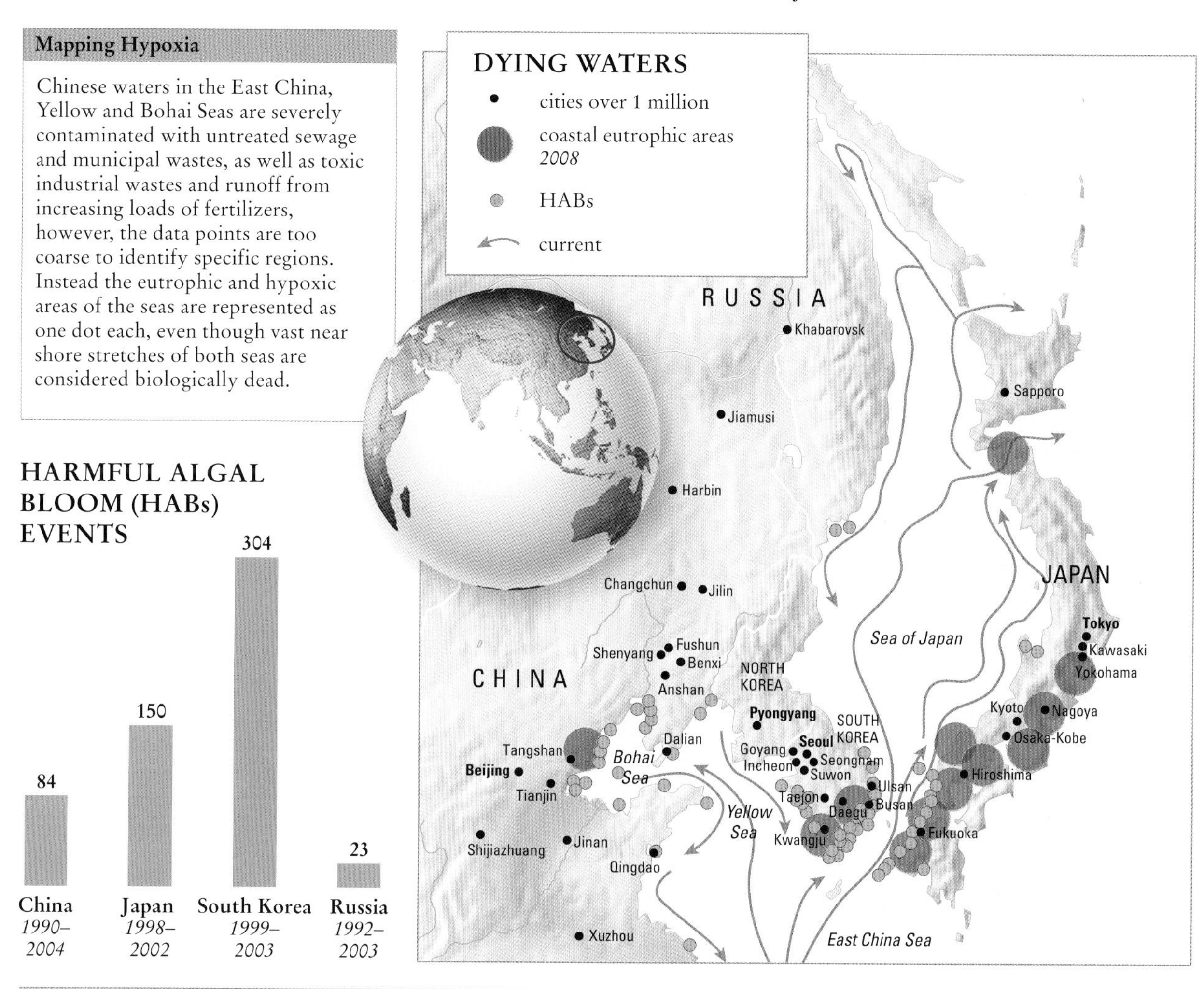

POLLUTED RIVERS

High population densities, poor wastewater treatment by countries bordering the seas (except in Japan), and high industrial outputs result in rivers carrying toxic waste into these seas. Biological oxygen demand is a measure of the oxygen used in decomposing organic wastes, such as raw sewage and agricultural runoffs. Chemical oxygen demand is a measure of the oxygen required to break down pollutants such as industrial wastes.

FROM RIVERS INTO COASTAL WATERS

2002

	river water inflows (km³)	BOD (tons/year)	COD (tons/year)
China	193 km³	582,874	1,012,665
Japan	125 km³	126,197	364,187
South Korea	46 km³	129,137	272,683
Russia	43 km³	84,410	438,821

Key Coastal Environments at Risk

Coastal wetlands, estuaries, and seagrass meadows perform vital ecological functions on which human safety and well being depends.

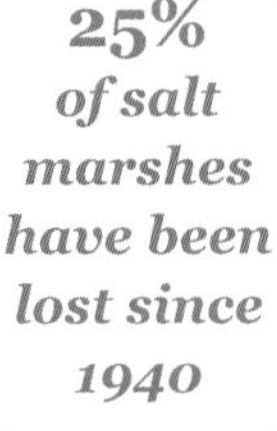

Coastal ecosystems are the intersection between land and sea, keeping life in the ocean healthy and protecting the land and human communities from the force of the ocean. However, the concentration of human populations along coastlines is putting these protective habitats at risk.

Estuaries and associated coastal wetlands – salt marshes, swamps, mud flats, and mangrove forests, along with seagrass communities – are among the planet's most fecund ecosystems. Hectare for hectare, they sustain more wildlife, both in numbers and diversity, and more primary plant growth than any other habitats on earth. These diverse and productive natural systems provide vital spawning, nursery, and feeding grounds for thousands of species of fish and shellfish, filter out pollutants washed off the land (such as heavy metals and nitrates), trap and stabilize sediments, serve as buffers between land and sea, modify climate, absorb flood waters, dissipate storm surges, and cycle nutrients from the land to near-shore ecosystems and the ocean.

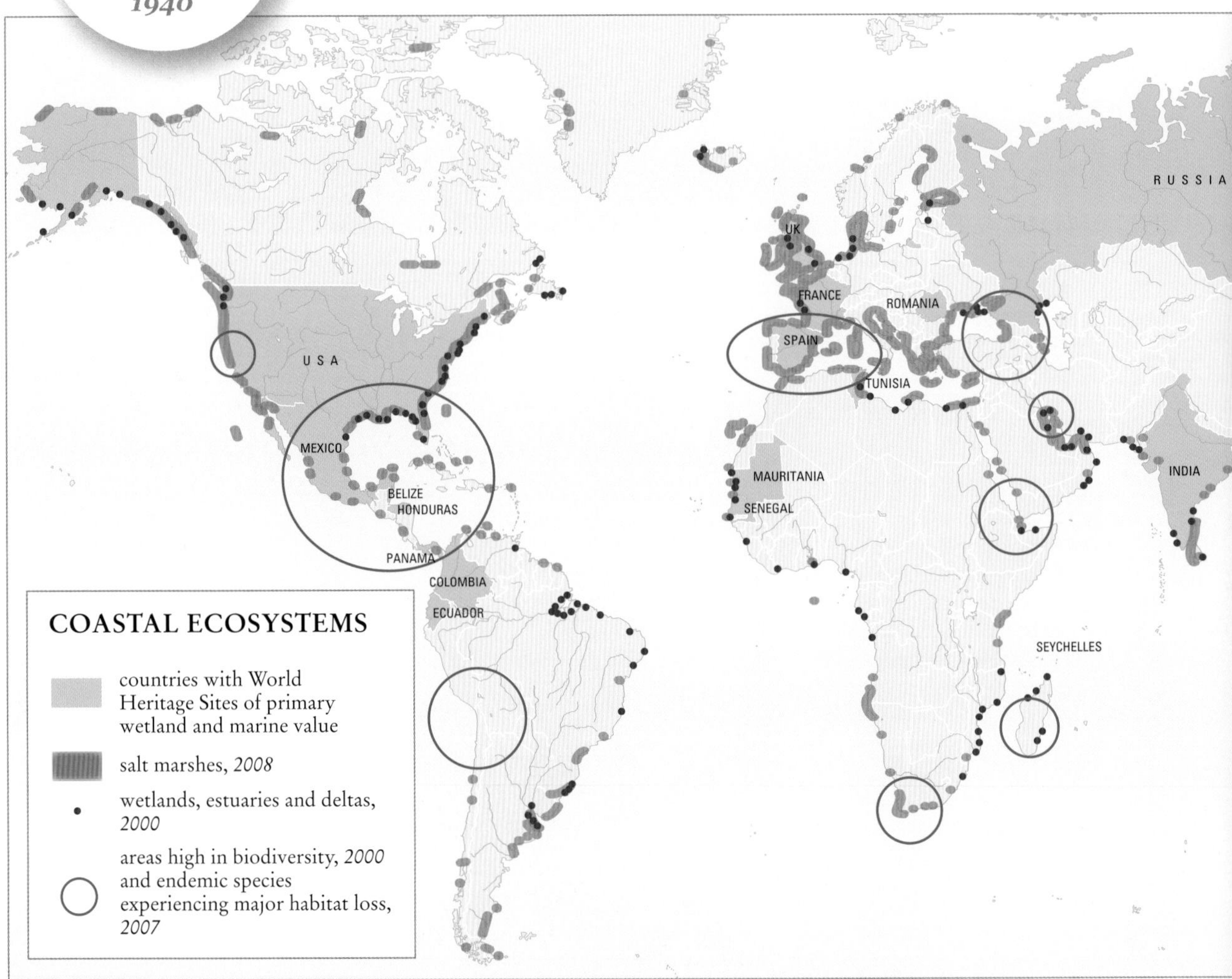

24–25, 26–27

Vegetated margins, in particular salt marshes, mangroves, and seagrasses, comprise less than 0.5 percent of the world's ocean area, but contribute about 50 percent of the total organic carbon holding capacity of the ocean sediments. They thus qualify as the most effective carbon sinks in the biosphere, exceeding the carbon binding capacity of Amazonian rainforests.

About one-quarter of the area originally covered by salt marshes and ponds has been lost, most of it in the past 60 years. Over the same period, the world has lost close to 30 percent of mangroves and one-third of global seagrass area. In other words, roughly one-third of the area covered by these valuable carbon sinks has been lost already, with the remainder threatened. Marine vegetated habitats rank among the most threatened ecosystems on the planet, with global loss rates estimated to be 2 to 15 times greater than the rate of loss of tropical forests.

VALUABLE ECOSYSTEMS

Estimated annual economic value of Australia's coastal ecosystems
$AUS per km²
2005

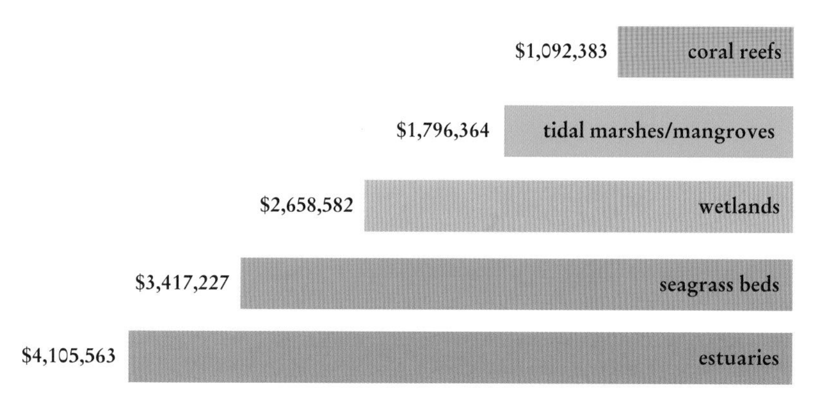

Blue carbon sinks

Marine plants play an important role in absorbing carbon dioxide, and therefore help to slow global climate change. Salt marshes, mangroves, and seagrass beds lock away an estimated 1.6 billion tons of carbon every year, making them one of the most intensive carbon sinks on the planet.

44–45, 46–47

Environments at Risk: Seagrasses

Seagrasses are the only land plants to have returned to the sea. With the exception of the polar seas, these underwater meadows are found in nearly all seas, and cover some 600,000 square kilometers (230,000 square miles) of near-shore areas. All 60 species of seagrass grow close to the shore in shallow water. The coral triangle in South-East Asia is a centre for seagrass biodiversity. Some 40 percent of all known species are found in the South China Sea alone.

Like mangroves, seagrasses trap and consolidate sediment, provide nurseries, shelter, and food for a host of marine life, reduce wave energy, absorb carbon dioxide, and regulate water flow. As no-cost fish farms, these ecosystems are unrivalled. A study of coastal Florida revealed that seagrass communities harbor 100 species of fish and 30 species of crustaceans; the latter are a key component of seagrass food webs. Another study documented 450 algal species dependent on seagrass beds.

Like mangroves and coral reefs, seagrasses contribute directly to the economy of local communities. One recent study puts the value of seagrasses at $3,500 per hectare (2.5 acres), based on their value as habitats for commercial fisheries and shore protection. Another study carried out in Florida estimated the value of five commercial species of fish dependent on seagrass beds at $48.7 million a year, not including their value for recreational fisheries.

Yet seagrasses are under threat in most places where they are found, vulnerable to coastal pollution, dredging for ports and harbors, trawl fishing, aquaculture, beach development, and sea level rise.

Though 72 countries have areas of seagrasses under some form of protection, their current status is uncertain. More efforts are needed at documenting the current condition of seagrass communities and protecting areas of high biodiversity.

SEAGRASSES OVER TIME

Percentage of surveyed seagrass sites in each category

1879–2006

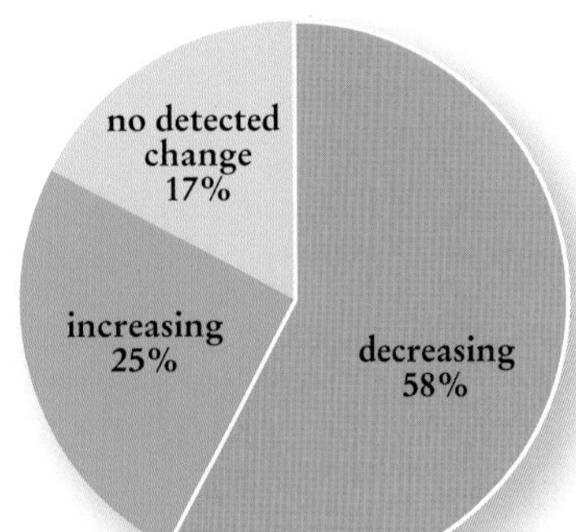

CAUSES OF HABITAT DESTRUCTION AND IMPACT ON ECOSYSTEM

Causes

- Land reclamation
- Sedimentation
- Land-based pollution (untreated municipal and industrial wastes, runoff from agricultural activities)
- Oil spills
- Coral mining
- Trawl fishing
- Transportation/navigation

Impacts

- Wholesale loss of biodiversity
- Loss of habitat for associated and dependant plants and animals, ranging from algae to the dugongs (endangered) and marine turtles that feed on seagrasses
- Loss of fisheries production as hundreds of species of fish use seagrass beds as breeding, nursery and feeding areas

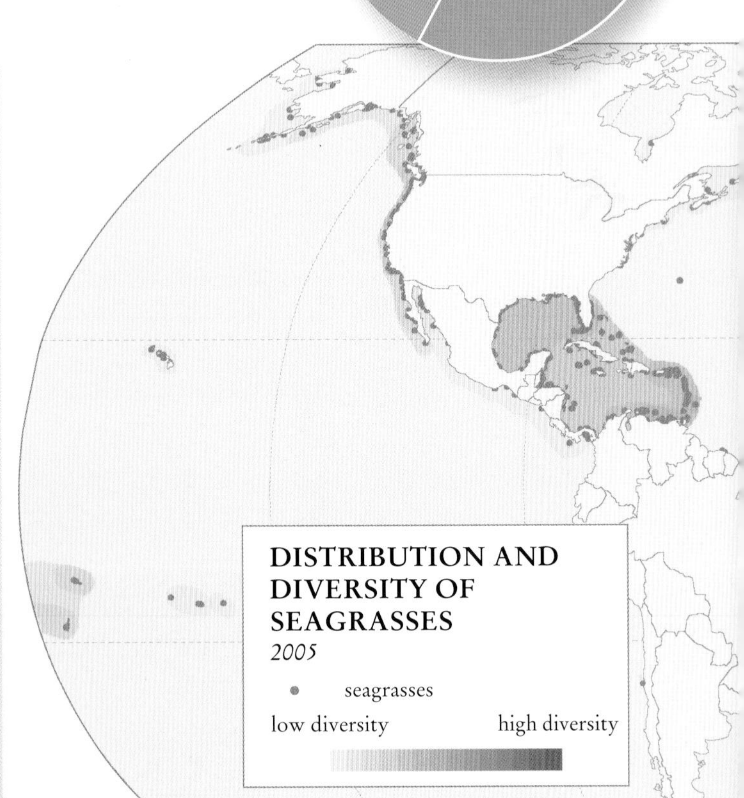

DISTRIBUTION AND DIVERSITY OF SEAGRASSES

2005

42–43

RISING RATE OF CHANGE

Number of sites surveyed showing change in area

1930–2000

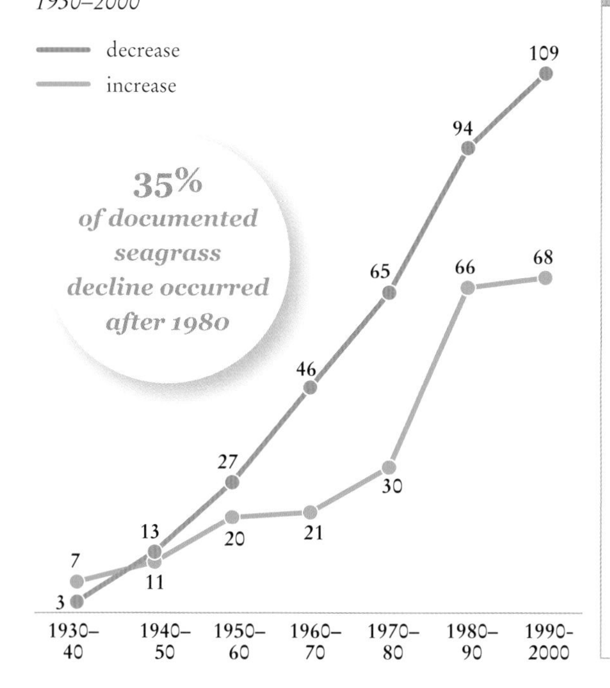

Seagrass Monitoring

Not all seagrass areas are decreasing – a 2009 study found 51 sites that were increasing. Of these, 11 were attributed to improved water quality and restored habitat. The report noted that transplantation efforts have generally failed, but watershed management and habitat remediation are effective.

Tampa Bay, Florida, is an example of this trend. Efforts to reduce nutrient runoff have resulted in a 50 percent improvement in water quality (as measured by clarity), and a recovery of 27 square kilometers (10.4 square miles) of seagrass beds in the last 25 years.

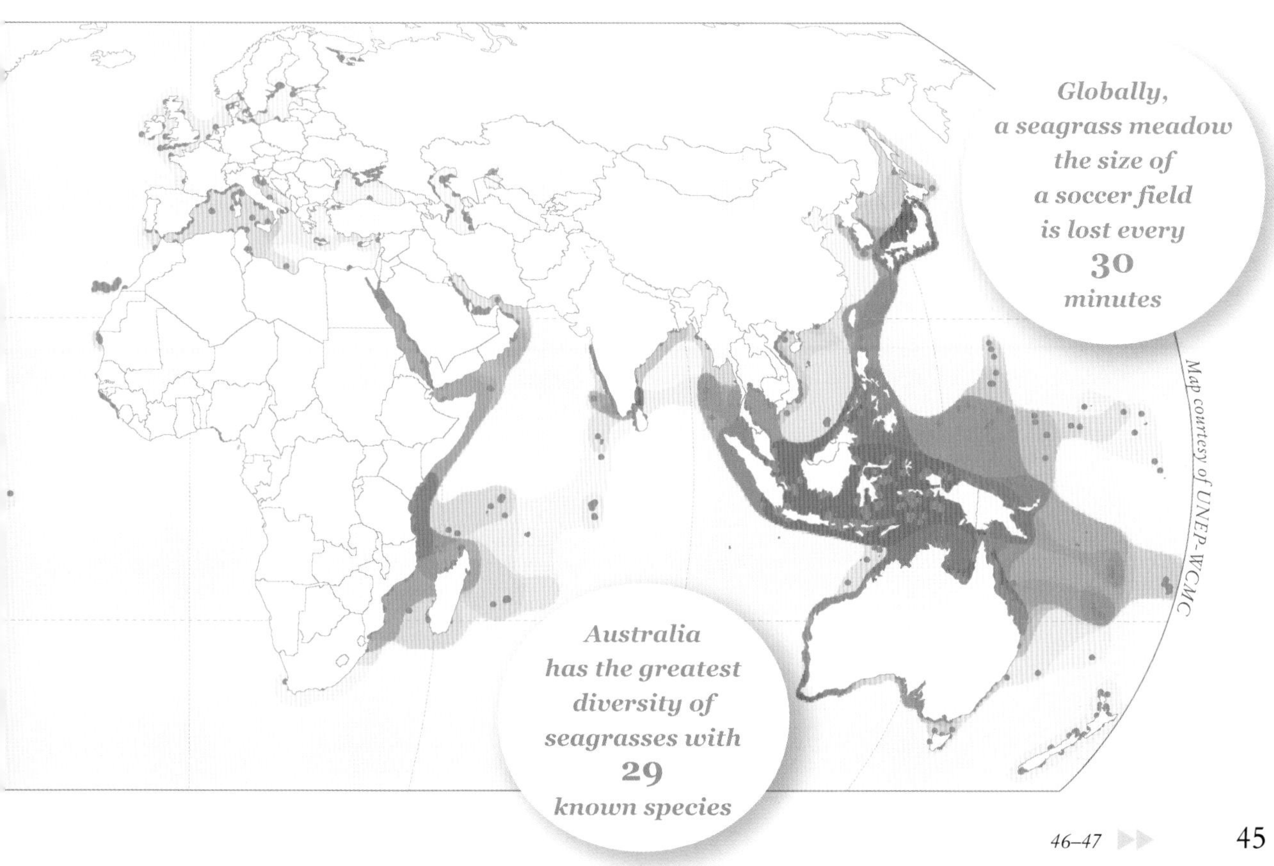

46–47

Environments at Risk: Mangroves

Mangroves are salt tolerant trees and shrubs found along sheltered coastlines in shallow-water lagoons, estuaries and river deltas in 124 tropical and sub-tropical countries. These unique forests are transition ecosystems between terrestrial and marine communities of plants and animals. They buffer tides, waves, and storm surges, protecting shorelines from the ravages of the sea and reducing erosion. They stabilize and trap sediments and nutrients washed off the land or ferried in by river systems, and their extensive root systems provide shelter and feeding areas for a wealth of marine life. There are some 60 species of mangroves, varying in height from a few meters up to an impressive 30 meters (100 feet).

If managed properly, mangroves provide numerous benefits to local economies. They are a source of wood (including fuelwood and charcoal), and non-wood products such as thatch, animal fodder, alcohol, sugar, medicines, and honey, as well as tannin suitable for tanning leather and curing fishing nets.

Mangroves also act as natural fish farms, providing breeding, feeding, and nursery habitats for commercially valuable species of fish and shellfish, including many species of shrimp and prawns. Assessments of the links between mangrove forests and local fisheries indicate that for every hectare of mangrove forest cleared, nearby coastal fisheries lose some 480 kg of fish and shellfish every year.

Mangroves are in decline throughout their range, victims of urban sprawl, tourism developments, expansion of agriculture, fuelwood collectors and charcoal makers, and conversion to fish and shrimp ponds. However, the rate of loss has decreased from 187,000 hectares lost during the 1980s, to 102,000 hectares lost during the period 2000–2005. Still, the overall picture is bleak: during the past quarter century some 3.6 million hectares have been lost, 20 percent of the world's total mangrove area as measured in 1980.

1980:* 18.8 *million hectares of mangroves* *2005:* 15.2 *million

Mangroves

Many mangroves have long aerial or air-breathing, above-surface, roots enabling them to withstand immersion in water and oxygen poor mud, which make it look as if the trees are propped up on stilts.

MANGROVES: DISTRIBUTION AND DIVERSITY
2005

mangroves

low diversity — high diversity

The longterm sequestration of carbon by* 1 *km²* *of mangroves is equivalent to that of* 50 *km² of tropical forest

 ◀◀ 28–29, 42–43, 44–45

Fish from trees

Around 80% of the Indian and Bangladeshi fish catches from the lower delta region of the Ganges, Brahmaputra, and Meghna rivers come from the extensive mangrove swamps of the Sunderban. These cover some 20,000 square kilometers (7,700 square miles) in the Bay of Bengal and are the largest mangrove bioregion in the world, spanning Bangladesh and India.

MANGROVE DECLINE

Extent of mangroves by region
1980–2005
million hectares

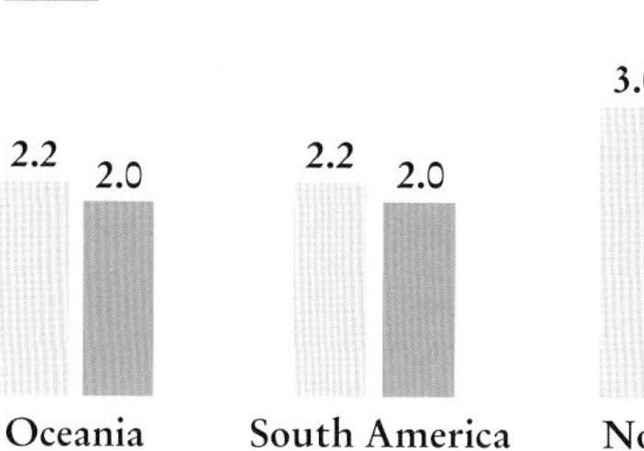

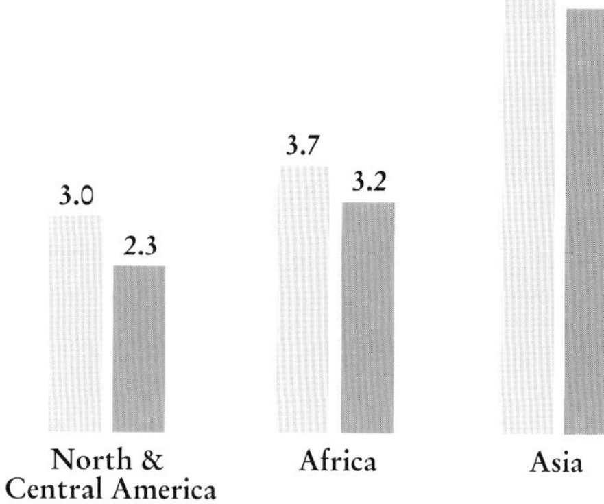

Mangroves: worth their salt

The worth of mangrove forests has been put at $200,000 – $900,000 per hectare per year, depending on the type of products and ecosystem services they provide. These vary from shore protection, to providing commercial fish stocks with breeding, nursery, and feeding grounds.

RETAINING MANGROVES

Estimated costs per hectare of different approaches in Thailand
2008

protecting healthy mangroves	rehabilitating degraded mangroves
$189	$946

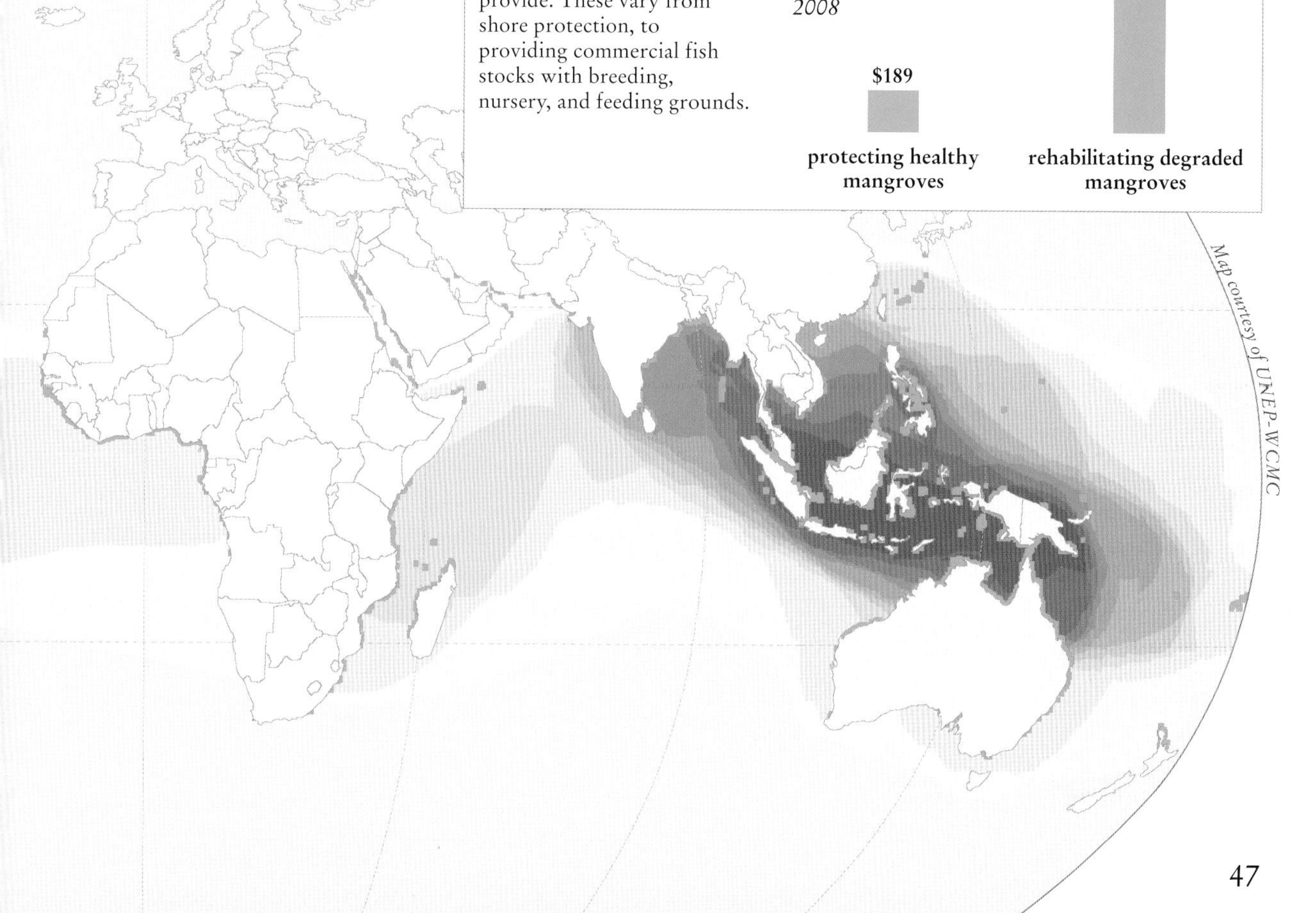

Map courtesy of UNEP-WCMC

Coral Reefs in Danger

Coral reefs provide valuable coastal protection, and play a vital role in the health and ecology of the oceans.

Warm-water coral reefs covering a total of nearly 285,000 square kilometers (110,000 square miles) are found in the coastal waters of some 100 tropical and sub-tropical countries. Around 500 million people benefit from the protection and the income they provide.

Coral reefs protect coastlines from storm damage and beach erosion, and provide homes, breeding areas, nurseries, and food for tens of thousands of species of fish, shellfish, and invertebrates. They also form an important link in cycling nutrients from the land to the open ocean, and are a potential pharmacopeia of new drugs to treat cancers, HIV, and other diseases. Although coral reefs occupy only one thousandth of the ocean's area, they are wonders of biological diversity – the rainforests of the sea. Marine scientists estimate that there may be a million species of plants and animals on tropical coral reefs.

Reef-based fisheries may contribute up to 80 percent of fish caught by subsistence or small-scale fishing communities, providing invaluable sources of food and protein. The needs of growing populations often conflict with the needs of these ecosystems, however. The area with the highest coral reef biodiversity – the so-called "coral triangle" – is under tremendous pressure from non-sustainable fishing practices, including the widespread use of dynamite and poisons to catch fish, coral mining for construction material and pollution runoff from the land.

The master builders of the sea

Coral reefs have existed for millions of years, but most present day reef structures were formed 10,000 years ago, after the last ice age. They are built by polyps, which are animals, consisting of a columnar body topped by stinging tentacles that fringe a central mouth.

Each polyp, which may be as small as a seed or as large as a lily pad, secretes calcium carbonate (limestone) that forms a cup in which it lives. Within polyp tissues live colonies of algae, which are the driving force behind their growth and productivity. In a symbiotic relationship, these tiny algae convert the sun's energy into food, provide oxygen for waste removal, and are responsible for some species' spectacular colors.

Most corals are colonial, creating masses of cups fused together in huge apartment-like complexes. The bulk of any reef is actually dead, only the upper layer being covered by living corals. Nevertheless, the entire structure supports myriad life forms and shelters two populations of marine organisms – day feeders and night feeders – who share the same residence in shifts.

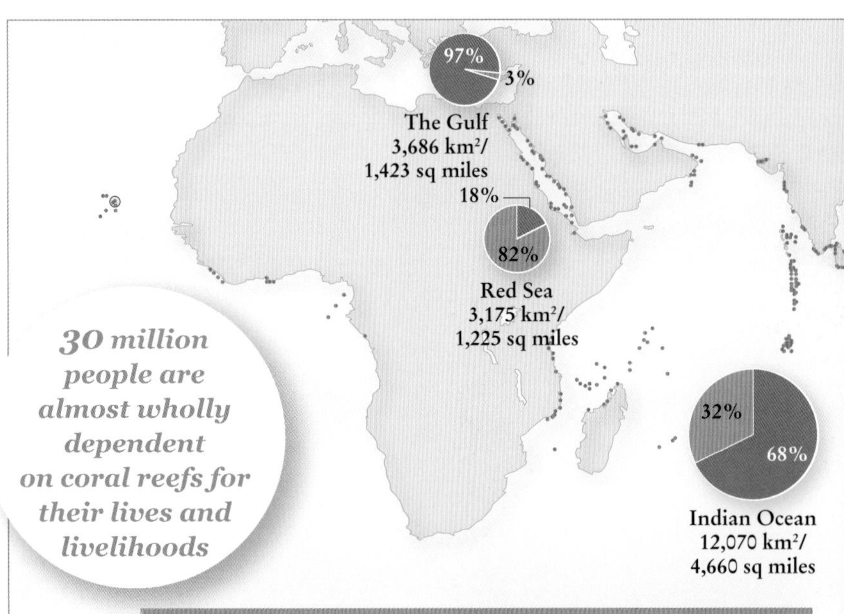

30 million people are almost wholly dependent on coral reefs for their lives and livelihoods

Coral bleaching

Coral bleaching is a threat to reefs. It can occur when water temperatures are raised by 1°C or more. The coral polyps expel their symbiotic algae, turning the corals white – a process that usually ends with the polyps' death.

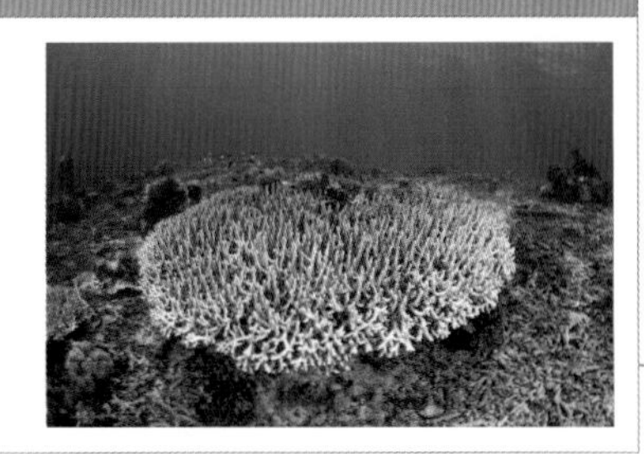

Human actions also have an impact in less direct, but no less damaging, ways. This includes coral bleaching, which occurs when the ocean temperature rises, and which could be exacerbated by climate change. The acidification of the oceans, recently identified as being the result of increased levels of atmospheric CO_2, may also make it increasingly difficult for polyps to build their limestone homes. Invasive species such, as the crown-of-thorns starfish, also pose an increasing hazard, with mass infestations thought to be triggered by lack of prey species and pollution from land-based sources.

Alarmingly, recent research carried out at the Australian Institute of Marine Science has indicated that climate-driven changes in ocean temperatures and chemistry could effectively kill off most hard coral species by mid-century, with unknown consequences for remaining coral communities.

THE FUTURE FOR CORAL

The state of coral worldwide

2008

Total: 284,800 km² / 110,000 sq miles

already lost	19%	
critical	15%	likely to be lost in 10–20 years
threatened	20%	could be gone in 20–40 years
healthy	46%	but could be affected by climate change

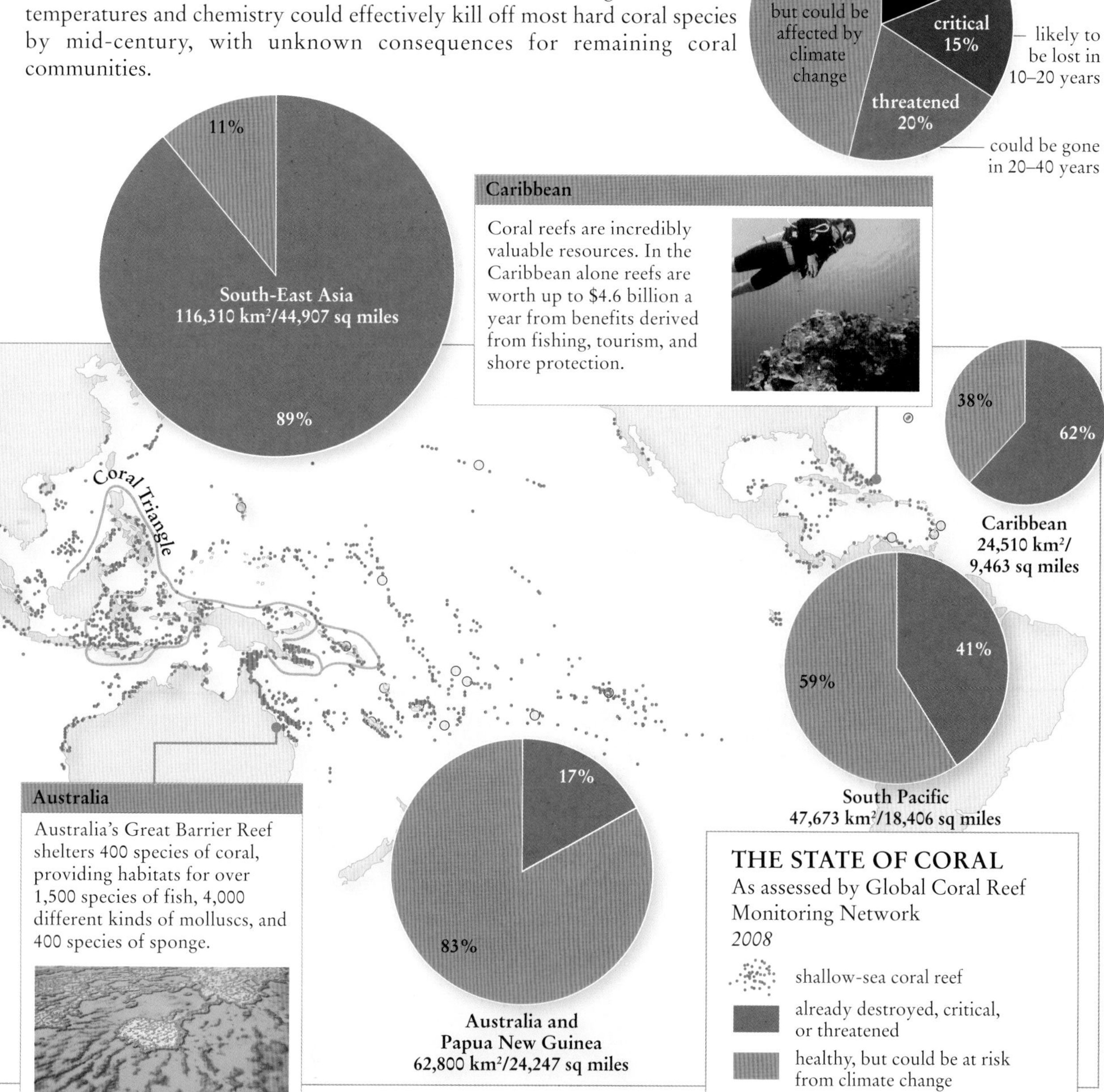

Caribbean

Coral reefs are incredibly valuable resources. In the Caribbean alone reefs are worth up to $4.6 billion a year from benefits derived from fishing, tourism, and shore protection.

Australia

Australia's Great Barrier Reef shelters 400 species of coral, providing habitats for over 1,500 species of fish, 4,000 different kinds of molluscs, and 400 species of sponge.

THE STATE OF CORAL

As assessed by Global Coral Reef Monitoring Network

2008

- shallow-sea coral reef
- already destroyed, critical, or threatened
- healthy, but could be at risk from climate change

80–81

The Empty Ocean

The rising global demand for seafood is leading to over-exploitation of fish stocks.

The prognosis for the health of the oceans and the vast fisheries they sustain is not good. In recent years one fishery after another has either collapsed commercially, or is being fished at its maximum yield. The state of wild ocean fisheries is so imperilled that if current overfishing continues, all the species we like to eat, and many more, will probably be gone by the middle of the century. The majestic blue fin tuna is already nearing the point at which it will be considered commercially extinct in the Mediterranean, and stocks are heading that way in the North Atlantic and elsewhere.

There are three main reasons for the demise of ocean fisheries: a precipitous rise in world demand for seafood, too many fishing fleets chasing too few fish, and the dismal failure of nearly all fisheries management efforts.

Not only are fishing fleets much better at locating and catching target species, but fishing gear has enabled trawlers to catch many species of no commercial value and leave a trail of total destruction through ocean-floor ecosystems. The offshore waters of developing countries in Africa and Asia are being plundered of all valuable fish and shellfish by deep-water fleets striving to supply the world's ever-increasing demand for seafood. This is

STATE OF TUNA
Number of species in each category
2007

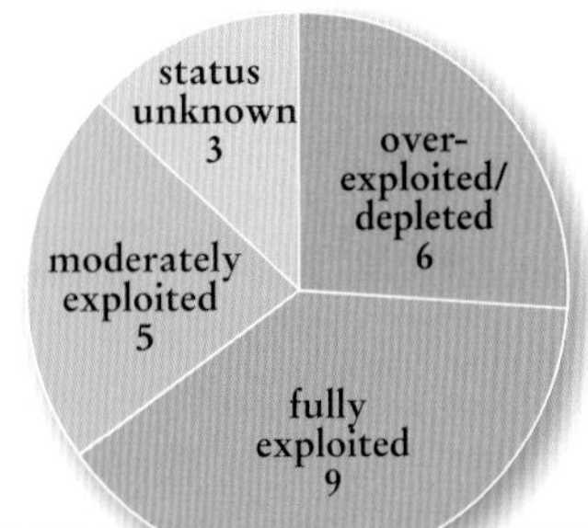

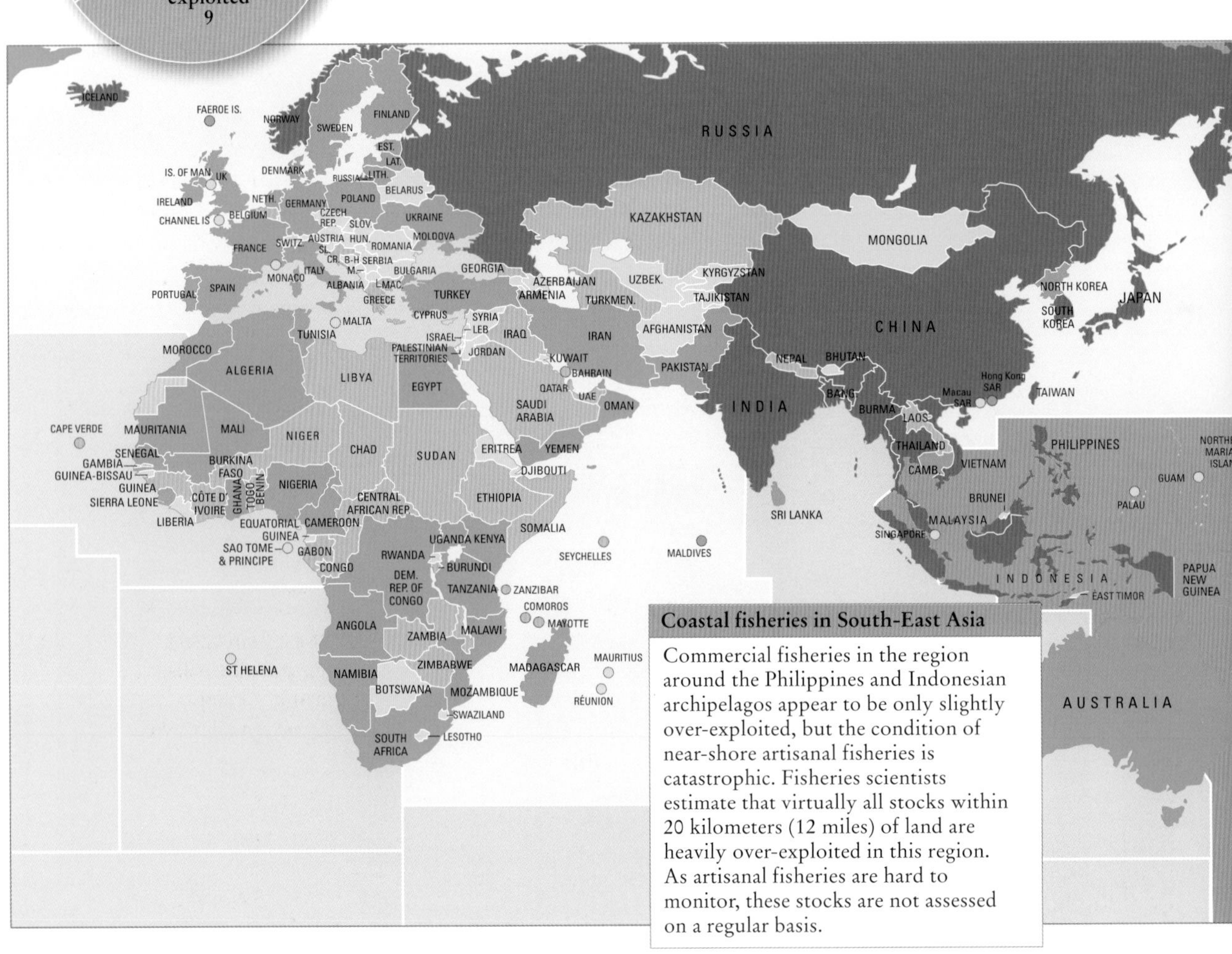

Coastal fisheries in South-East Asia

Commercial fisheries in the region around the Philippines and Indonesian archipelagos appear to be only slightly over-exploited, but the condition of near-shore artisanal fisheries is catastrophic. Fisheries scientists estimate that virtually all stocks within 20 kilometers (12 miles) of land are heavily over-exploited in this region. As artisanal fisheries are hard to monitor, these stocks are not assessed on a regular basis.

◀◀ 34–35

likely to end with the collapse of important fisheries and the loss of valuable sources of protein. The demise of the cod fishery on the Grand Banks off Canada should have been an early warning. When cod stocks collapsed in 1992 and the fishery had to be closed, scientists thought that stocks would recover in a couple of years. This had still not happened 18 years later.

Although demand for seafood is rising, only one-fifth of all commercially exploitable species assessed by the FAO present an opportunity for increased production. With some 2.6 billion people depending on seafood for at least 20 percent of their protein intake, the loss of fisheries has ominous implications for the future food security of poor coastal and near-shore populations.

LOSS OF STOCKS

Percentage of fish stocks collapsed

1950–2003, prediction to 2050

global fisheries data

extrapolated long-term trend

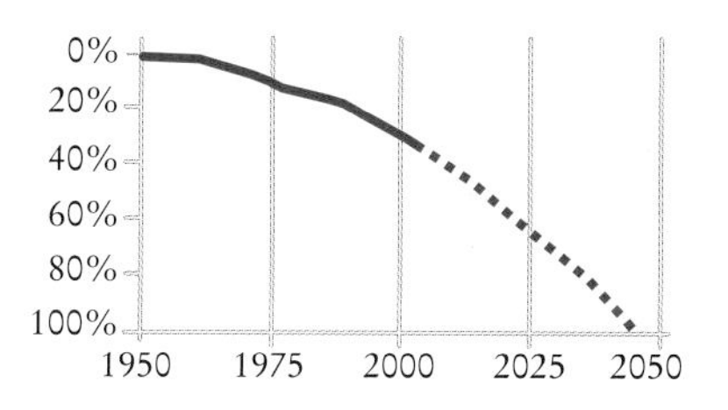

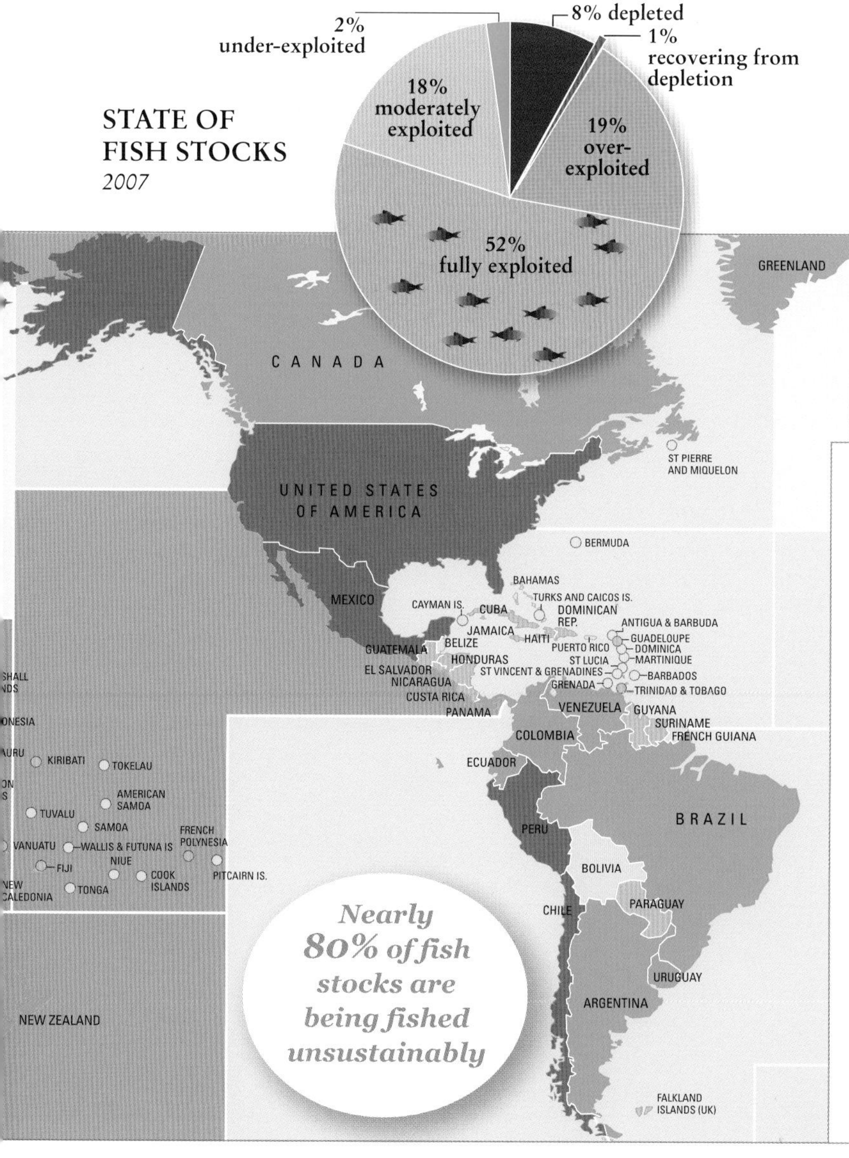

> “Unless we fundamentally change the way we manage all the ocean species together, as working ecosystems, then this century is the last century of wild seafood.”
>
> Dr. Steven Palumbi, Stanford University

FISH PRODUCTION

Total fish caught

2008

tons

- 1 million or more
- 100,000 – 999,999
- 10,000 – 99,999
- fewer than 10,000
- no data

Level of exploitation

Percentage of fish stocks exploited at or beyond maximum sustainable yield

2008

- 71% – 80%
- 20% – 52%
- 10% or less

Fishing boats in Morocco: Fishing provides a livelihood for many small-scale producers.

Part Three: Trade, Commerce and Tourism

Major Shipping Lanes

World trade is made possible by the oceans. The bulk of the world's products and goods travel by ship to key seaports, then onward by rail or road.

The overwhelming majority of world trade travels by ship. Some 90 percent of all products and goods – amounting to some 30 trillion ton-miles a year – are shipped by sea, a significant amount of it leaving or arriving at 18 major sea ports around the world. In all, there are some 45,000 merchant ships plying the world's oceans. The heaviest traffic is through the Panama and Suez Canals, the two major transit routes for shipping to and from Europe and Asia or between Asia and the Caribbean and the east coast of North America.

Shipping poses six major threats to the marine environment: 1) routine discharges of oily bilge water; 2) dumping of non-biodegradable solid waste into the ocean; 3) accidental spills of oil or toxic wastes while in port or underway; 4) emissions to the atmosphere from marine fuel burned by the ship's engines; 5) port and inland channel construction, including dredging of channels, turning basins and berths for ships (which often means the filling-in of nearby wetlands); and 6) ecological harm due to the introduction of exotic species transported in vessel's ballast water (considered by IUCN to be one of the major global threats to biodiversity).

GLOBAL SHIPPING ROUTES

2004

- most heavily used routes
- medium to heavy use
- light to medium use
- lightly used

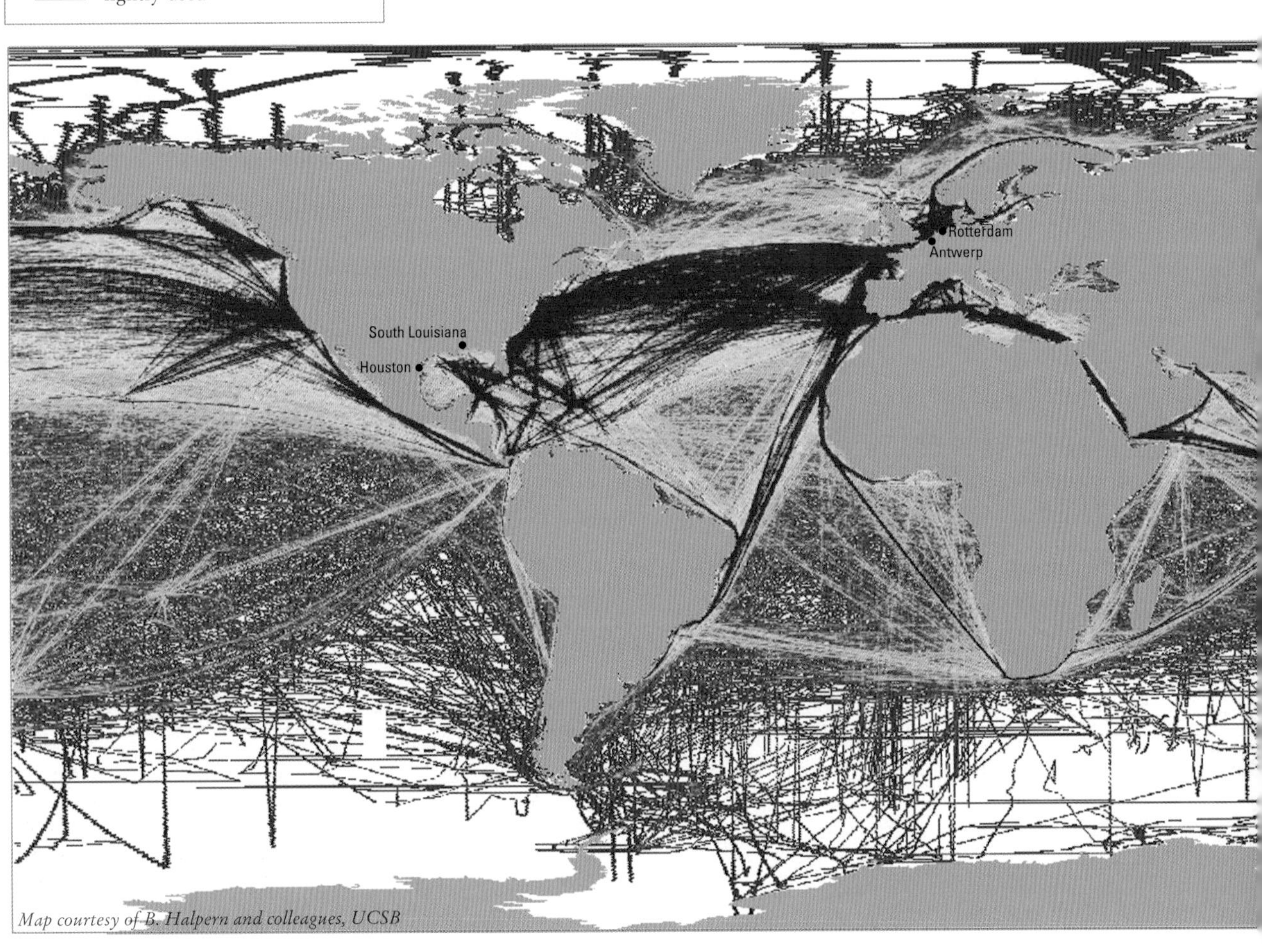

Map courtesy of B. Halpern and colleagues, UCSB

34–35

Oil tankers and bulk carriers often run empty one way, taking on large quantities of ballast water in the process. This is one reason for the proliferation of exotic or invasive species – they can hitch-hike in ballast waters, which are discharged at the end of the outgoing or incoming route. Invasive species cost the US alone around $120 billion a year in damaged ecosystems.

Air pollution from ship's engines remains a problem, both for the environment and human health. Shipping is responsible for 18 to 30 percent of the world's nitrogen oxide pollution and about nine percent of global sulphur dioxide pollution. The US has imposed a 230-mile buffer zone along its coasts in an effort to reduce the health costs associated with shipping, estimated at some $330 billion a year due to lung and heart diseases.

85% of all ship pollution occurs in the northern hemisphere

Shipping is responsible for up to 4% of climate change emissions

AMERICAN ASSOCIATION OF PORT AUTHORITIES TOP TWENTY PORTS
By total cargo volume
2008

1. Singapore, Singapore
2. Shanghai, China
3. Rotterdam, Netherlands
4. Tianjin, China
5. Ningbo, China
6. Guangzhou, China
7. Qingdao, China
8. Hong Kong, China
9. Qinhuangdao, China
10. Dalian, China
11. Busan, South Korea
12. Nagoya, Japan
13. Shenzhen, China
14. South Louisiana, US
15. Houston, US
16. Antwerp, Belgium
17. Ulsan, South Korea
18. Chiba, Japan
19. Port Hedland, Australia
20. Port Kelang, Malaysia

Climate change opens fabled Northeast Passage

For centuries, mariners have searched in vain for an ice-free passage through the Arctic Ocean to connect Europe and North America with Asia. Many died in the attempt. In August 2008, NASA satellite pictures showed that for the first time in 125,000 years, both the Northeast and Northwest Passages were open at the same time.

With higher temperatures at the top of the world as a result of climate change, there is now a real possibility of an ice-free corridor through the Arctic on a regular basis. In September 2009, two German freighters entered Rotterdam Harbor, after navigating the Arctic coast of Russia from Vladivostok in the Far East. The month long journey shaved 10 days off the usual 17,700 kilometer (11,000 mile) route through the Indian Ocean, saving each ship over $300,000 in transport costs.

This was only possible because the Arctic icecap is retreating at an alarming rate, leaving vast swaths of open water where solid pack ice recently frustrated attempts at even summer navigation.

By late this century, Russian climatologists predict that the navigable period through Arctic Russia might grow to several months from the 6–8 week ice-free window the passage now offers each summer. Already, shipping companies are looking at the Russian route as a viable way to save time and money while ferrying goods from East to West.

Energy from the Sea: Oil and Gas

The quantity of oil extracted from offshore fields is expected to increase as land-based reserves are exhausted, posing an increasing hazard to the marine environment.

The world still runs on hydrocarbons – oil, gas, and coal. In 2008, global production was 82 million barrels of oil per day and 3 trillion cubic meters of gas. At present, 70 percent of this comes from land based sources and 30 percent from offshore reserves. Global consumption is expected to increase by one percent per year, and, with most land-based fields identified and exploited, companies are looking to untapped offshore reserves to meet this increasing demand.

In August 2010, a UK company, Cairn Energy, found gas in Baffin Bay, off the west coast of Greenland, prompting speculation that oil deposits would be discovered in the next few years. A rush for Arctic hydrocarbons and minerals is expected as climate change reduces sea ice cover.

Currently there are more than 900 offshore oil and gas fields in Europe's North Sea. The UK accounts for 486 of them, Norway for most of the rest. The European Commission, fearing fallout from a possible oil disaster in the North Sea similar to the Deepwater Horizon blowout in the Gulf of Mexico during the summer of 2010, wants to ensure that energy firms drilling in European waters can cover the costs of a catastrophic spill. New rules on offshore exploitation of hydrocarbons are expected in 2011.

Diffuse sources of hydrocarbon pollution create low but persistent contamination over large areas of ocean. Many aspects of chemical

OFFSHORE OIL RESERVES
2009

- offshore hydrocarbons exploited or under development
- offshore reserves in exploratory phase
- no coastline, or offshore reserves not yet explored

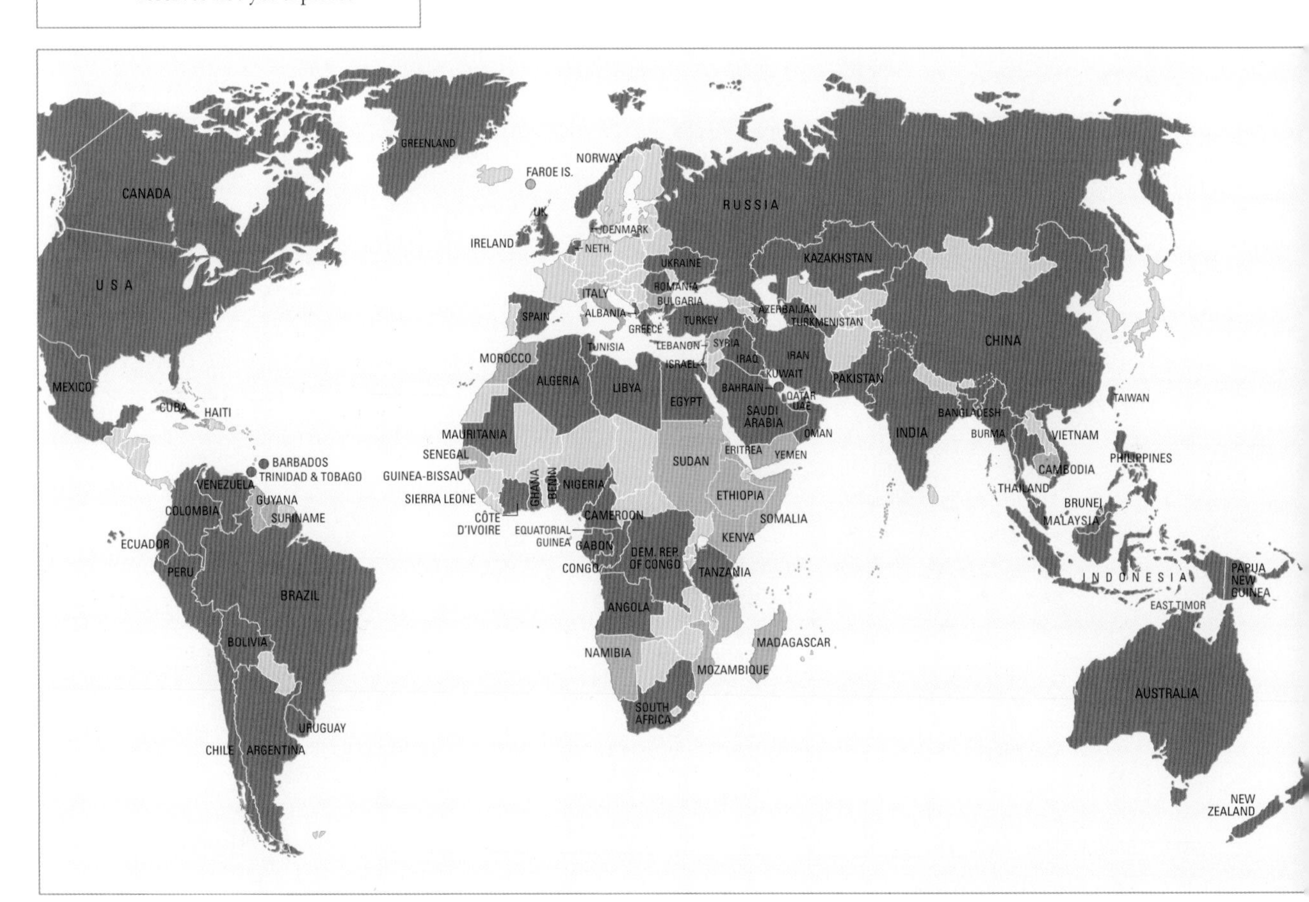

34–45

composition of hydrocarbon pollution and its biological impacts remain poorly understood. Studies have revealed, however, that those areas of highest biological productivity: estuaries, and coastal and shelf waters, have the highest levels of pollution from oil and related contaminants, nearly all of it from land-based sources, such as oily contaminants washed off urban streets and parking lots, and industrial activities.

Oil pollution from offshore fields is falling as a result of both tighter environmental controls and technological advances. In the North Sea, for instance, improvements in controlling and treating drilling fluids and mud have reduced oil contamination of surrounding waters by 50 percent over the last 15 years.

SOURCES OF OIL POLLUTION ENTERING THE OCEANS

2002
million tons

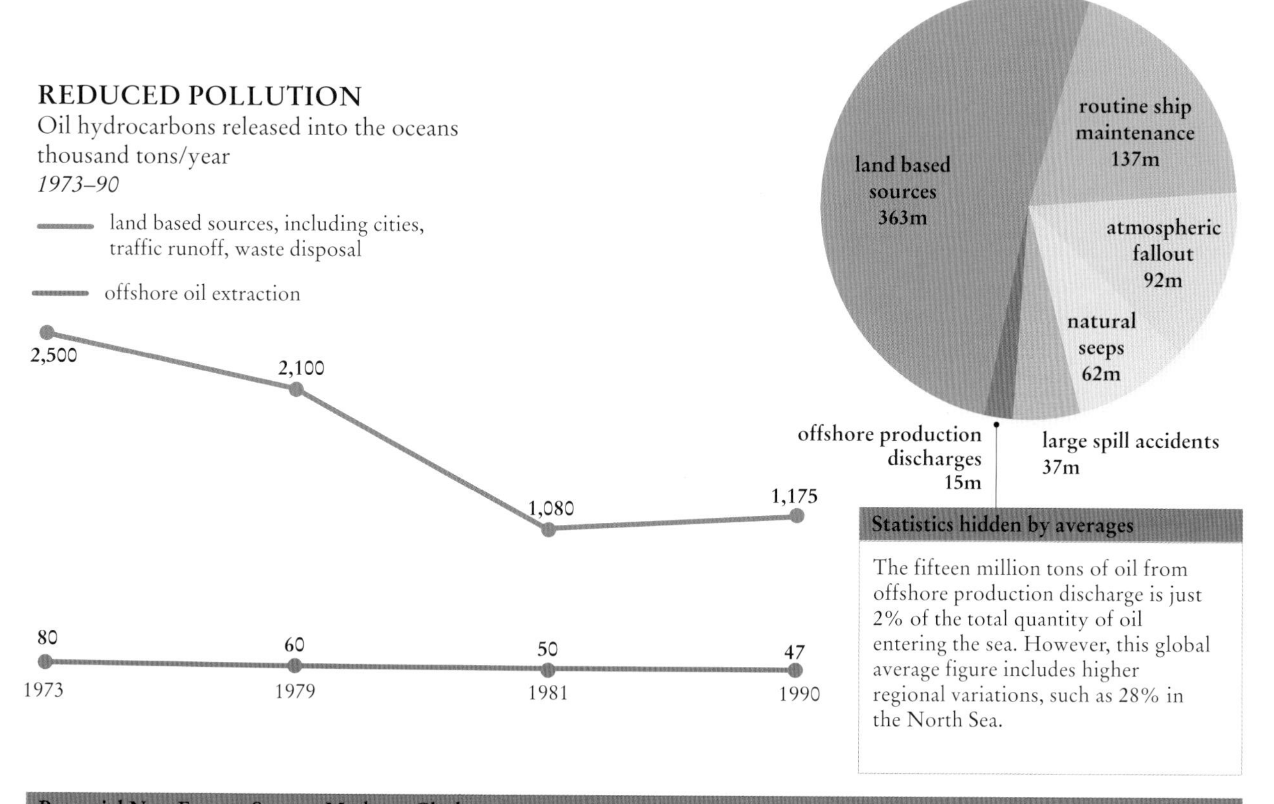

Statistics hidden by averages

The fifteen million tons of oil from offshore production discharge is just 2% of the total quantity of oil entering the sea. However, this global average figure includes higher regional variations, such as 28% in the North Sea.

Potential New Energy Source: Methane Clathrates

Methane clathrates are essentially methane gas trapped inside ice crystals, occurring mainly in the Arctic permafrosts and on the ocean's continental shelves in cold, highly pressurized environments. Preliminary research suggests that they may comprise a significant portion of total fossil fuel reserves.

Clathrates are known to be explosively unstable, especially if the temperature increases or the pressure decreases – as a result of climate change, tectonic uplift, or undersea landslides. With conventional supplies of fossil fuels becoming depleted, there is growing interest in the possibility of using methane clathrates as an energy source. Exploratory work is currently underway in the US and Canada to determine the viability of extracting and utilizing pressurized methane clathrates.

Methane is a much more powerful climate-changing gas than carbon dioxide. Some scientists have expressed concerns that the process of extracting the methane clathrates could cause them to break in an uncontrollable chain reaction that could release large quantities of the gas into the atmosphere, in turn accelerating climate change.

58–59, 60–61

Energy from the sea: Wind

By mid-2008, wind power was generating over 100,000 megawatts (MW) of electricity in more than 70 countries. The overwhelming majority of these installations are on land. However, offshore wind farms, once thought too expensive to build and operate, are gaining in number and importance and will supply a significant share of wind-driven electricity in Europe and elsewhere by 2030.

By October 2010, according to the Global Offshore Wind Farm Database, 52 sea-based wind farms were operating in 11 countries, generating 3,000 MW of electricity. A further 650 wind farms in 36 countries were in the planning stage, permitted to advance, or under construction.

The UK leads the world in installed offshore capacity, with an impressive 13 offshore wind farms generating 1,341 MW of electricity. The UK is followed closely by Denmark with 863 MW of installed capacity and the Netherlands with 247 MW. These three countries, along with Germany, have the best offshore wind conditions in Europe to support large arrays of wind turbines.

Europe's offshore potential is enormous. The European Environment Agency (EEA) estimates the potential of offshore wind generated electricity at around 25 million MW by 2020; seven times greater than projected electricity demand in Europe.

The USA is planning some 88 offshore wind farms capable of generating more than 35,000 MW of electricity. The majority of these however are still on the drawing board.

One of the delays in expanding offshore capacity is red tape. In a number of countries firms planning offshore wind farms are hampered by restrictive regulations, complicated licensing processes and lengthy approval times.

So far, major environmental impacts have not been detected with offshore wind installations. Bird kills and interference with radar installations have been flagged as potential concerns, but studies are inconclusive. The biggest environmental problems for land-based systems – visual pollution and noise – are not factors in offshore sites. The other factor which makes offshore wind an attractive alternative to land-based wind farms is that each unit can be built much larger, in the range of 2 to 5 MW or even bigger.

By 2015 offshore wind will provide ***20%*** *of Europe's wind power capacity*

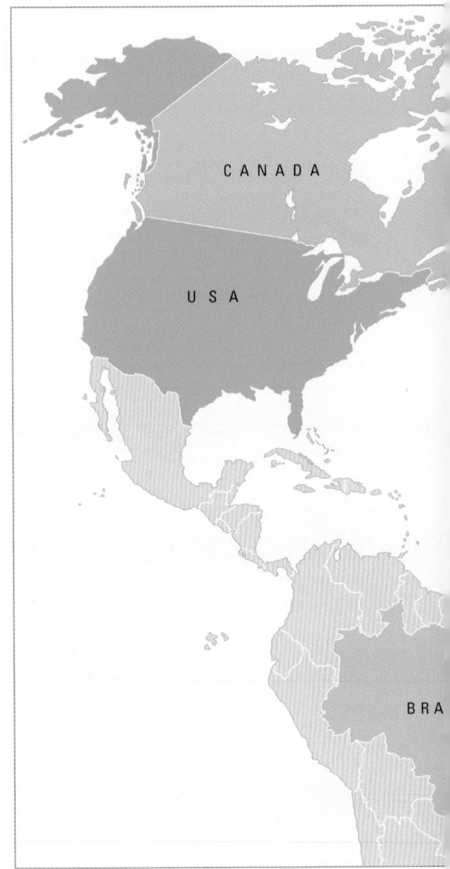

Offshore wind turbines in Denmark.

56–57

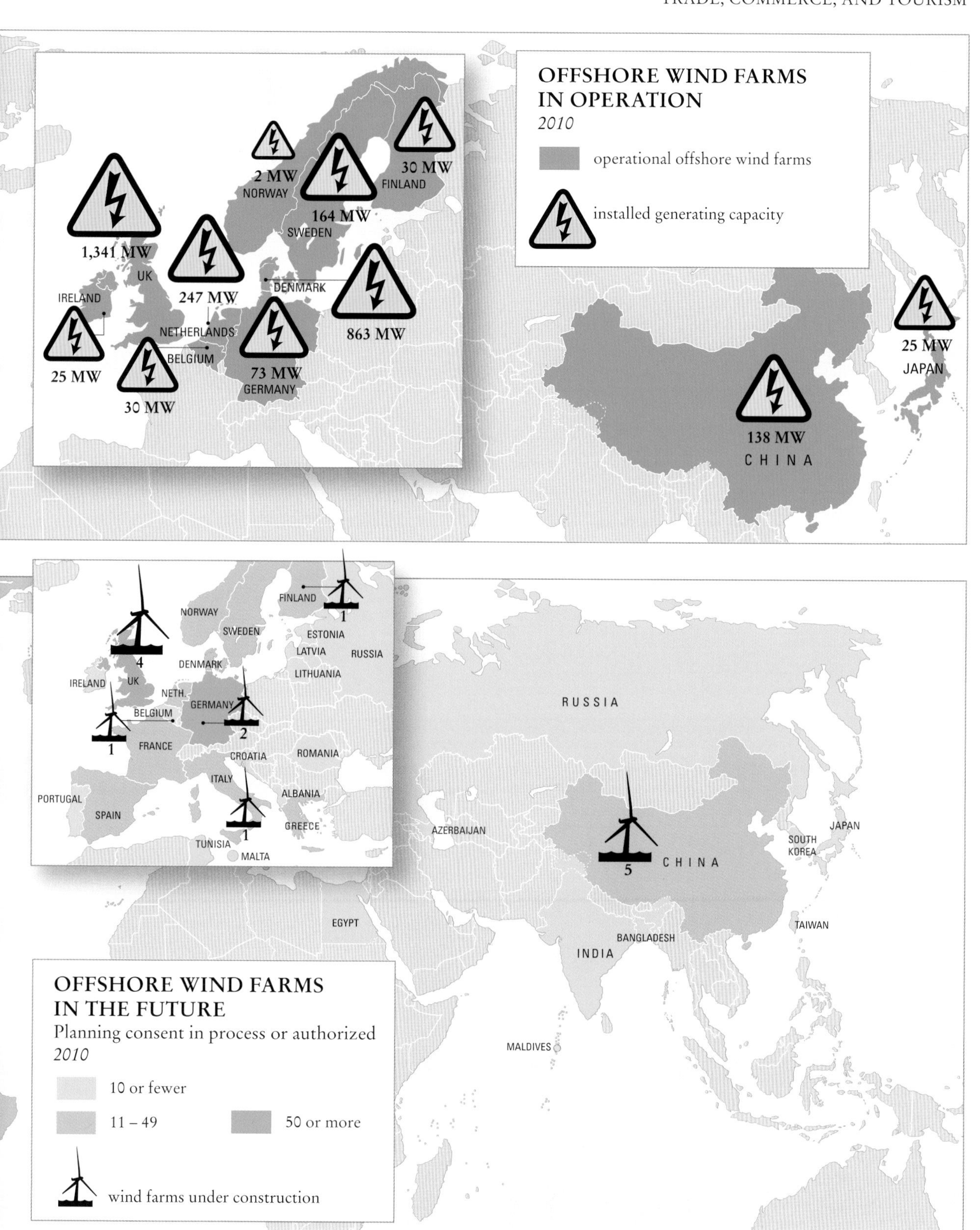

OFFSHORE WIND FARMS IN OPERATION
2010
operational offshore wind farms
installed generating capacity
2 MW
NORWAY
30 MW
FINLAND
164 MW
SWEDEN
1,341 MW
UK
IRELAND
247 MW
DENMARK
863 MW
NETHERLANDS
BELGIUM
25 MW
30 MW
73 MW
GERMANY
25 MW
JAPAN
138 MW
CHINA
OFFSHORE WIND FARMS IN THE FUTURE
Planning consent in process or authorized
2010
10 or fewer
11 – 49
50 or more
wind farms under construction
FINLAND
1
NORWAY
SWEDEN
ESTONIA
LATVIA
RUSSIA
4
DENMARK
LITHUANIA
IRELAND
UK
NETH.
GERMANY
BELGIUM
2
1
FRANCE
CROATIA
ROMANIA
ITALY
ALBANIA
PORTUGAL
SPAIN
GREECE
1
TUNISIA
MALTA
RUSSIA
AZERBAIJAN
5
CHINA
SOUTH KOREA
JAPAN
EGYPT
TAIWAN
BANGLADESH
INDIA
MALDIVES

Energy from the Sea: Tides and Waves

The world's oceans are vast and so is their potential to generate power from tides, waves, and temperature gradients. According to the European Energy Association the global potential of energy from the oceans is around 100,000 TWh (terawatt hours) per year; dwarfing total world electricity consumption, which in 2009 stood at about 16,000 TWh a year.

Given its enormous energy potential, scientists have estimated that, with current technologies under development or being tested, the oceans could produce between 8,000 and 80,000 TWh of electricity a year from wave energy, 2,200 TWh from tidal currents and 10,000 TWh from ocean thermal (temperature) gradients. At the beginning of 2009, some 24 countries were deploying, developing, and testing a variety of ocean renewable technologies.

The power of waves breaking on the world's coastlines has been estimated at 2–3 million MW

Currently there are five distinct types of renewable energy that can be harnessed from the oceans. *Tidal energy*, derived from the force of tides is harnessed by building fixed or floating barrages in estuaries or near shore waters. *Tidal current*, which exploits the kinetic energy present in currents can be harnessed by using modular underwater turbine systems which rotate in the current like underwater wind turbines. *Wave energy*, currently the most developed of the systems, taps the potential energy created by the ebb and flow of waves based on floating or submerged systems. *Temperature gradient*, also known as Ocean Thermal Energy Conversion (OTEC) exploits the temperature differences between warmer surface water and deeper colder water. *Salinity gradient* systems can be placed in river deltas to take advantage of the difference between fresh and salt water gradients, using an osmotic process (similar to those used by desalination plants which turn salt water into freshwater).

None of these options has as yet emerged as the clear leader in harnessing potential energy from the ocean, but all are currently undergoing trials or being expanded.

Canada and US poised to dive into ocean energy

The US Pacific Northwest coast has the most potential for developing wave and tidal power. Scientists estimate that this region could produce 40 to 79 kilowatts of electricity per meter of coastline. Wave energy alone could provide up to 6.5 percent of US electricity demand.

Canada, which has the world's longest coastline – some 200,000 kilometers (124,200 miles) – is developing both tidal and wave energy systems. The potential energy to be derived from 190 test sites off the north, west and east coasts could provide a full two-thirds of the country's electricity demands.

UK ocean energy projects

The UK has the largest number of ocean energy projects and companies developing various technologies: nearly 40 projects in 2008, with some 300 companies active in the field. The UK plans to build wave and tidal energy installations capable of generating 1.2 GW of electricity, enough to power 750,000 homes, around the Orkney Islands and in the Pentland Firth, off the north coast of Scotland. Six sites have been allocated for wave energy development, generating 600 MW, with four tidal projects planned, also generating 600 MW.

◀◀ 56–57

TIDAL ENERGY

Tidal barrage plants have been in operation for decades. The oldest tidal energy plant was built in La Rance, Brittany, in France and has been in continuous operation for over 40 years, with a maximum generating capacity of 240 MW of electricity. Currently, South Korea is constructing the world's largest tidal power station, which will generate 260 MW of electricity when in operation; an even larger one generating 520 MW is expected to go on-stream in 2014.

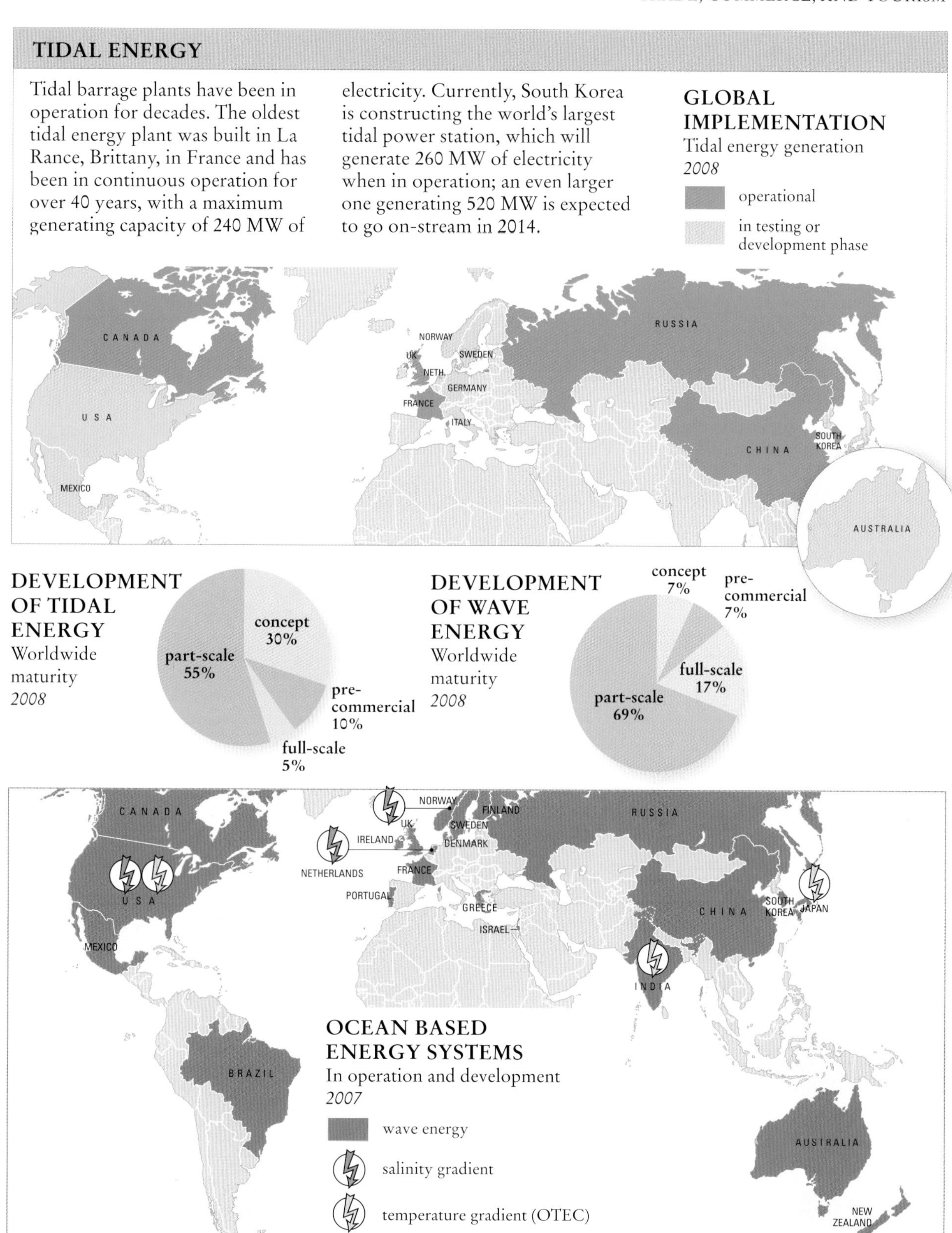

Coastal and Marine Tourism

Globally tourism has grown by 9 percent each year since the mid-1980s, with coastal and marine tourism the fastest growing sector of the travel and leisure industry.

The number of international tourists has doubled over the past 20 years, from 438 million in 1990 to 880 million in 2009; these numbers are expected to double again by 2020, reaching 1.6 billion. Collectively, international tourists spent $852 billion in 2009, much of it in areas offering sun, sea, and sand.

In the USA, for instance, the number of visitors to California's beaches averages around 567 million a year, more than the total number of visitors to all of the country's national parks combined.

The Caribbean has experienced dramatic increases in tourist arrivals over the past three decades. In 2009, nearly 19 million tourists descended on the region; an additional 18.1 million arrived on cruise ships, accounting for a full 36 percent of the region's GDP.

The land clearance and urbanization that flows from marine tourism have severe environmental impacts. Developers routinely clear littoral forests, rip out mangrove stands, dredge through seagrass meadows, and fill in wetlands to create tourism infrastructure, including hotels, resorts, restaurants, and marinas. Furthermore, solid and liquid wastes, especially sewage from resorts, hotels, and municipalities, are often pumped into coastal waters with little or no treatment.

MAJOR TOURIST DESTINATIONS
Numbers of international arrivals
2009

- 20 million or more
- 10 – 19.9 million
- 5 – 9.9 million
- 1 – 4.9 million
- fewer than 1 million
- no data

Tourism is the largest sector of the global economy accounting for 5% of global GDP

24–25, 26–27, 42–43, 44–45, 46–47

The explosive growth of packaged holidays on cruise ships, though contributing little to local economies other than host ports, is triggering environmental repercussions. At the beginning of 2008 there were over 200 large cruise ships in service, half of them in the Caribbean. By 2000, Caribbean cruise ships were generating around 90,000 tons of solid waste each year, much of it ending up in the sea or adjacent landfills.

Pollution from a variety of municipal and agricultural wastes has already damaged coral reefs in the Caribbean – one quarter have already been lost, while up to 90 percent in some regions are threatened, especially in the high impact tourism destinations – Jamaica, US Virgin Islands, Bahamas, and Puerto Rico. Marine scientists are concerned that other high biodiversity areas, including the coral triangle in South-East Asia, may end up similarly impoverished as a result of tourism development.

INCREASING TOURISM

Number of tourists by region
1995–2020 projected
millions

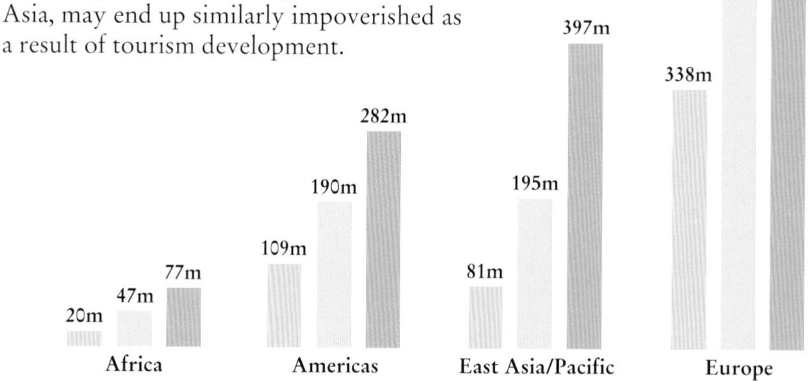

IMPACT OF TOURISTS ON COASTAL ECOSYSTEMS

(excluding infrastructure impacts)

Activity	Impact	Ecosystems
Sunbathing, picnicking	Litter, plant damage, fire hazards, stress to animals such as sea turtles	Sandy beaches, dunes
Swimming	Water contamination from sun-tan products, and other chemicals	Coastal waters, lagoons
Surfing, sailing, paddling	Movement, stress to animal species such as seals and water birds	Coastal waters, sea, beaches
Scuba diving, snorkelling, and underwater spear fishing	Damage to corals, loss of fish species, sediment disruption, water clouding and decreased photosynthesis, stress to fish	Coral reefs
Motorized water sports (motor boats, water skiing, jet skis, parasailing, sightseeing in glass-bottomed boats)	Noise, stress to animals, vibration, sediment disruption, damage to shore and underwater vegetation, petrol and oil contamination, injury, death, and/or poisoning of animals	Coastal waters, lagoons, river mouths, coral reefs
Fishing, clam diving	Over fishing, collection of particularly attractive species	Open sea, coastal waters, lagoons, river mouths, beaches

Coastal and Marine Tourism: Mediterranean Sea

The Mediterranean region is at the crossroads of three continents – Europe, Africa, and Asia. It has the most urbanized coastline of any sea; over half of its 46,000 kilometers are classified as urbanized or built on. The Italian coast is perhaps the most developed: over 43 percent is completely developed, as a result of urbanization, holiday homes, and tourism. Less than 29 percent is free from construction. Currently, two out of every three inhabitants of the Mediterranean live in urban areas and demographic growth, coupled with rural-to-urban migration, is driving urbanization trends across the entire basin, especially on the southern and eastern rims. With the addition of 275 million tourists per year, the coastline is under tremendous development pressure.

The Mediterranean is one of the world's 25 hot spots of biodiversity, despite the fact that the sea constitutes only 0.8 percent of the world's ocean area and 0.3 percent of its volume. It contains 7 percent of all known marine species; over 12,000 have been identified and described so far. A full 30 percent of the region's species are endemic, found nowhere else on the planet.

One of the Mediterranean's key marine ecosystems is its rich *Posidonia* seagrass beds, which grow close to shore. These rich ecosystems support a full 25 percent of all marine species but cover only 1.5 percent of its near-shore seabed, some 35,000 square kilometers. They provide spawning and nursery areas for commercially important fish, oxygenate the water, trap and fix sediment, and protect beaches from erosion.

The Mediterranean is losing its biodiversity – close to 20 percent of the region's species are endangered. In all, 63 percent of the sea's fish and 60 percent of its mammals have endangered status. This is attributed to the wholesale disappearance of coastal lagoons and seagrass beds, coastal erosion, over-exploitation of marine resources, and the expansion of invasive species, including unwanted aquarium species dumped into the sea.

The Mediterranean's future is uncertain. Though all littoral states are members of UNEP's Mediterranean Action Plan, more concerted efforts are needed to save ecosystems, husband resources, and manage urbanization and tourism.

14% of the Mediterranean coast is seriously damaged. **0.78%** is protected

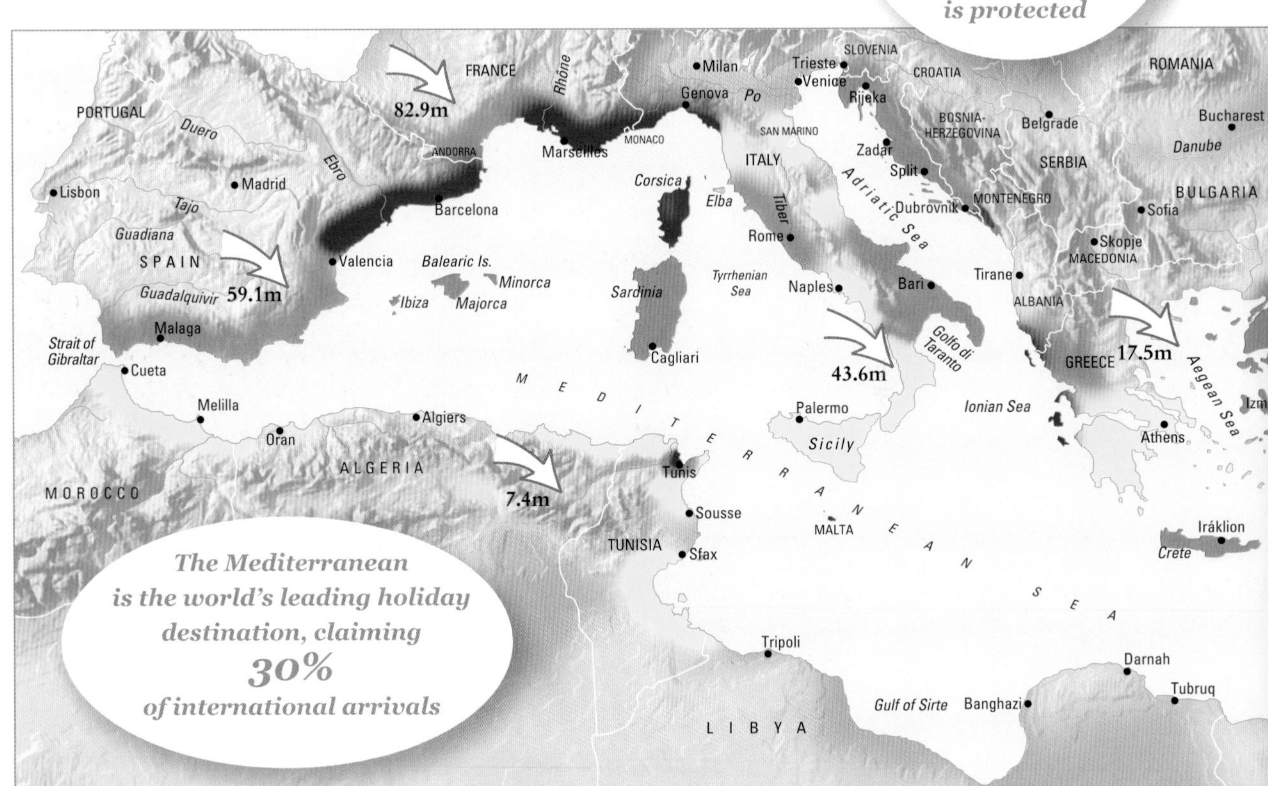

24–25, 26–27, 44–45, 62–63

The demise of Mediterranean fisheries

Currently, commercial fisheries land between 1.5 and 1.7 million tons of seafood per year. But catches do not satisfy regional demand – one-third of fish consumed in the Mediterranean are imported from other seas, most notably the Atlantic. Reckless and uncontrolled fishing during the 1980s is responsible for the collapse of numerous commercially valuable species such as tuna and squid. Catches have dropped by 25 percent.

Endangered species in the Mediterranean

Populations of monk seals, and the loggerhead and green turtle are declining alarmingly due to human pressure, habitat reduction, and accidental fishery by-catch. Only 150–250 monk seals remain in the region, mainly in the Ionian and Aegean seas, along the southern Turkish coast, and in some small groups in Morocco, Algeria, and Libya. Only one population of the loggerhead marine turtle remains and there are yet smaller numbers of the green sea turtle limited to the eastern parts of Turkey and Cyprus.

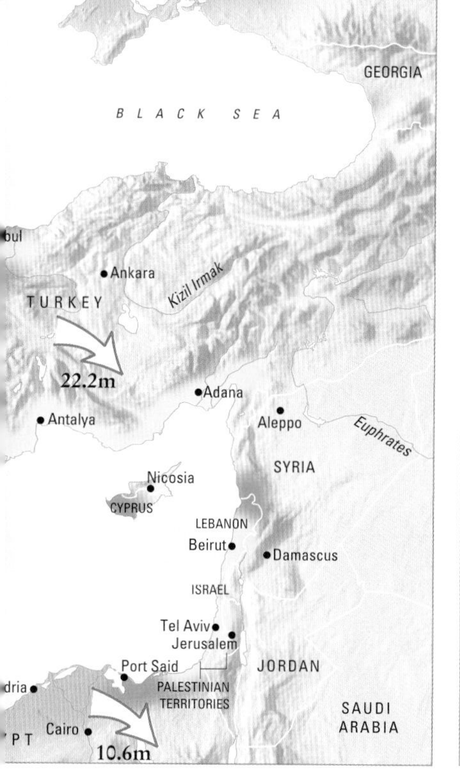

WASTEWATER BURDENS

The Mediterranean Sea is on the receiving end of some 10 billion tons of industrial and municipal wastes per year; much of it with little or no treatment. A staggering 40 percent of the region's municipalities – encompassing some 14 million people – have no access to wastewater treatment plants (673 cities out of 1,699).

DIRTY WATERS

Numbers of coastal cities without wastewater treatment plants
2004
selected countries

Spain	4
Italy	8
Greece	9
Syria	12
Algeria	58

Tourism and recreation account for **52%** *of coastal litter*

96–97, 98–99

Farming the Sea

Farmed fish and shellfish are meeting most of the rising demand for seafood from the developing economies of Asia, the Middle East, and Latin America.

Rapidly rising demand for fish and shellfish, both freshwater and saltwater varieties, has primarily been met by a huge leap in the quantity and value of farmed fish and shellfish on the world market: in 2006 it was worth nearly $80 billion. The output of aquaculture (freshwater) and mariculture (saltwater) operations has more than doubled since 1990, and, in the face of depleted wild fish stocks, now represents more than a third of all fish production, and provides almost half of the fish eaten by humans. (Roughly 75 percent of fish caught or produced are used for human consumption, 25 percent of the capture fishery goes to producing feed for fish farms, including the high proportion of by-catch that is too small for human consumption).

Aquaculture and mariculture operations range from simple backyard ponds, using natural food sources, to industrial-scale farming covering thousands of hectares of ponds and near-shore waters with highly intensive feeding, aeration, and water-quality control systems. Some large operations, such as Norway's Atlantic salmon farms, use massive marine pens, anchored in fjords.

Many freshwater farming operations are found close to or along coastal areas, as well as in riverine environments. Runoff and pollution from these operations impacts on the quality of coastal waters.

Crab fattening in Capiz Province, Philippines is just one aspect of the country's widespread small-scale mariculture operations.

WORLD MARICULTURE INCREASE

million tons

2000–08

fish and shellfish

aquatic plants, e.g.: seaweed

	2000	2002	2004	2006	2008
Top of bar	13.1	15.0	16.7	18.6	19.7
Lower segment	9.3	10.5	12.6	13.9	15.7

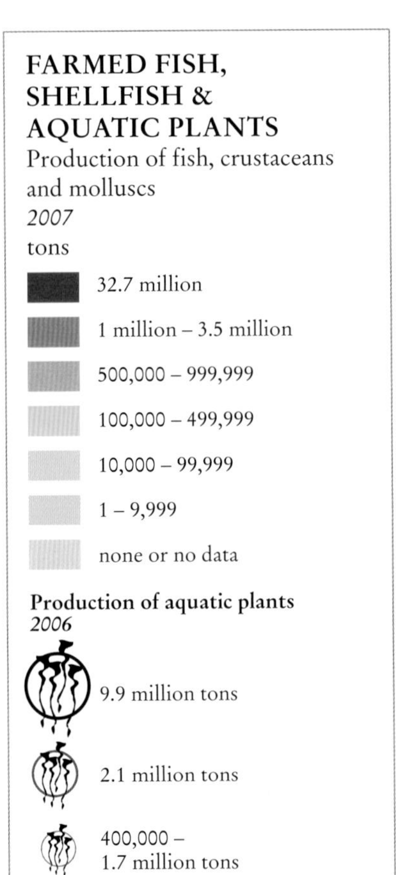

50–51

Around 220 species of fish and shellfish are farmed commercially. The Asia-Pacific region dominates the market, producing 98 percent of all carp, 95 percent of all oysters, and 88 percent of all shrimp and prawns. China is by far the dominant country, with 4.5 million fish and shellfish farmers producing two-thirds of all farmed fish, shellfish, and aquatic plants consumed globally.

The environmental impact of large-scale fish and shellfish farms has been devastating, with coastal wetlands – salt marshes and ponds, salt flats, and mangrove forests – being destroyed to make way for them. Ironically, this has adversely affected wild fish stocks. For every kilogram of shrimp farmed in Thai shrimp ponds, 400 grams of wild fish and shrimp are estimated to have been lost to those whose livelihoods depend on fishing. Fish farming makes other demands on wild fish stocks – wild fish are processed to create feed for carnivorous cage-raised fish, such as salmon.

The complete lack of coastal management plans in many developing countries has contributed to the rush to develop new fish and shellfish pond cultures. The wastes from these operations – uneaten feed, feces, and chemicals – unless disposed of properly, contribute to coastal and near-shore pollution, degrading water quality and harming wild stocks.

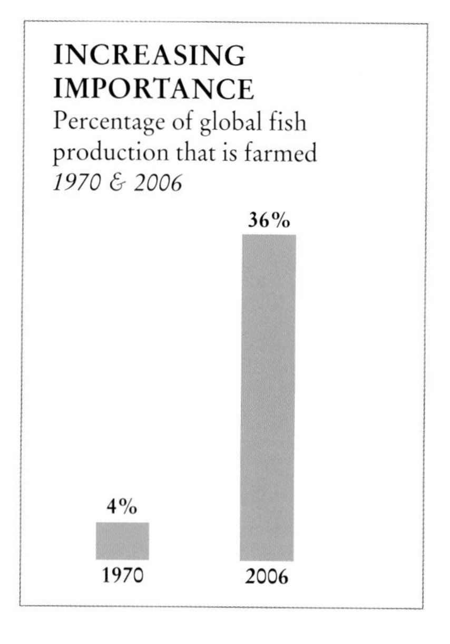

68–69 ▶▶

Farming the Sea: Asia and Indo-Pacific

Aquaculture and mariculture have grown in importance over the past two decades as capture fisheries continue to decline. While farming fish and shellfish has benefits, it has also created both social and environmental problems. In the Philippines, for example, mangroves, tidal creeks, and estuaries – habitats that traditionally provided a rich variety of seafood to small-scale fishers – have been cleared to create privately owned fish-ponds, pens, and cages.

The creation of shrimp ponds, in particular, has led to the loss of 70 percent to 80 percent of Vietnamese and Philippine mangrove forests, leading to impoverished coastal environments, with reduced catches of fish and shellfish, and diminished coastal protection and flood control. As long as the considerable benefits of maintaining mangroves and seagrasses are not widely understood, there will be little incentive for poor communities to maintain these valuable ecosystems.

The Asia-Pacific region is responsible for 91% of global aquaculture production

Many fish farming operations use wild-caught brood fish stock that is then matured, rather than hatched in tanks. In the Philippines, approximately 1.7 billion wild milkfish fry are captured annually to stock ponds. This removes valuable stock from coastal and ocean fisheries, reducing potential catches for capture fisheries.

Tank, cage, and pen fisheries have problems with disease, both from close confinement of many fish and from the introduction of exotic species. Escaped fish and discharges from mariculture operations can spread disease to wild populations. In an effort to control disease in fish farms, antibiotics are administered. These can lead to antibiotic resistance and contamination of sediments.

Outflows and sediment from fish farms, containing high nitrogen, phosphorus, and BOD levels, along with other pollutants can cause red tides, eutrophication, and may impact on the health of phytoplankton communities.

Small-scale fish farming, however, has significant benefits for poor communities across South-East Asia. Integrated coastal management could potentially ameliorate many of the negative impacts if the siting of farms, effluent management, and disease control are addressed in comprehensive management plans.

Seaweed farming in the Philippines

Blast fishing, a widespread fishing technique using dynamite to stun or kill fish, has been practiced for over 50 years across parts of Asia. Though blast fishermen can bring in more fish in the short term, the practice devastates reef and marine environments.

In an effort to curb the practice, in 2001, fisheries officials in the local government unit of San Lorenzo introduced seaweed farming on Guimaras Island as a potential alternative. The trial scheme began with four farms, and expanded to a further 19 villages. By 2006 there were 17 hectares under cultivation, with 162 farmers participating. In 2005, 6 tons of fresh and 22 tons of dried seaweed were sold by their traders association, at an estimated value of $14,977. Farmers involved have reported that the income derived from seaweed farming has been sufficient to keep them away from illegal fishing.

Seaweed farming is also less environmentally damaging than fish farming, and can provide shelter and habitat sites for many species.

AQUACULTURE SECTORS

Asia-Pacific region

2004

Total: 46.9 million tons

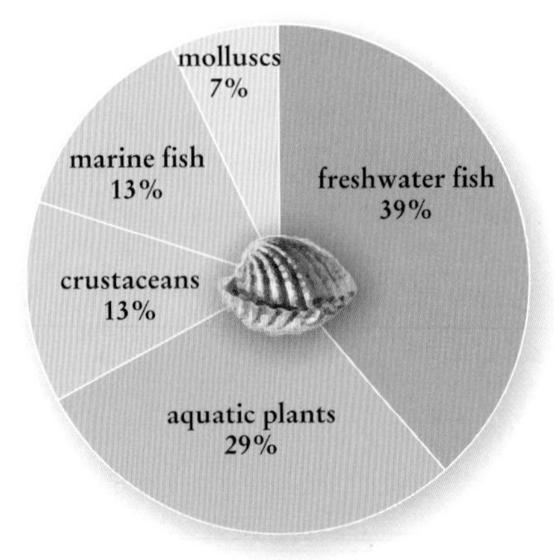

36–37, 40–41, 66–67

EMPLOYMENT IN AQUACULTURE

Number employed, not including temporary, occasional, or seasonal fish farmers

1995–2006

Small-scale aquaculture

Over 80 percent of aquaculture producers in Asia are small-scale farmers. Low start-up costs mean that it is possible for producers of very modest means to engage in crab-fattening, dried fish production, and farming of molluscs or seaweed to produce high-value foods.

ECONOMIC IMPORTANCE OF AQUACULTURE

Percentage of GDP contributed by aquaculture in Asia and Indo-Pacific region

2006

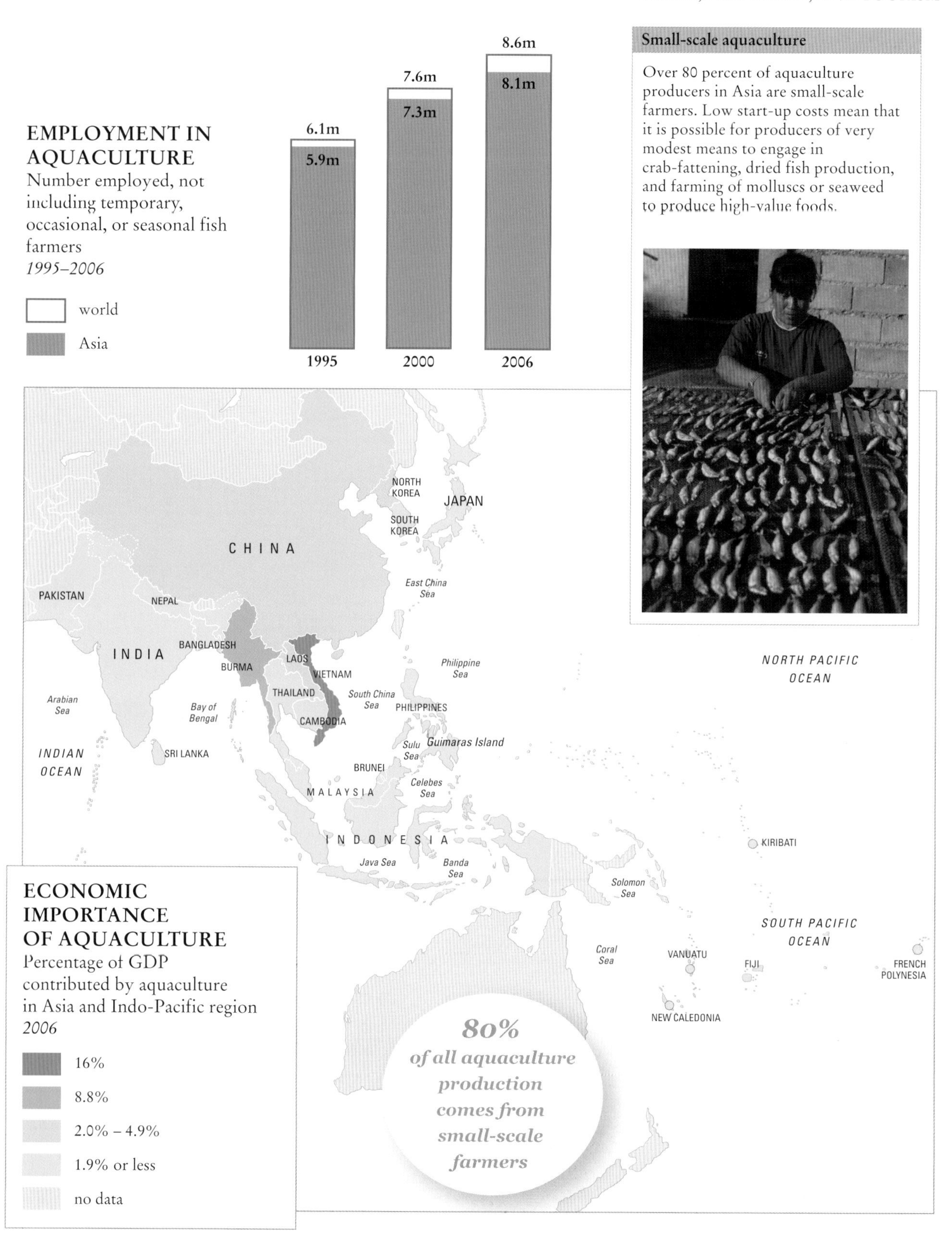

Storm waves break over the church in the Mediterranean town of Camogli in Italy.

Part Four: Climate Change

The Ocean Conveyor Belt

The world's oceans play an important role in shaping the earth's climate by circulating tremendous amounts of solar energy as warm water from the Pacific Ocean to the Atlantic.

The world's oceans are in constant motion. This motion is driven by thermohaline circulation (THC), from the Greek words *thermo*, meaning temperature, and *haline*, referring to the sea's salt content; these two factors determine the density of sea water. Warm water from the Pacific is taken to the Atlantic Ocean as a shallow near surface, mainly wind-driven, current. In its progress around Europe the current cools, releasing heat into the atmosphere and evaporating, becoming more salty. It finally mixes with cold water from the Arctic and becomes so cold, salty, and dense that it sinks and is returned to the Pacific as a deepwater current. Once back in the northern Pacific the water up-wells, returning to the surface where it is warmed and conveyed back to the Atlantic.

Considerable mixing takes place in the ocean basins. Some of the water up-wells in the Southern Ocean, but the oldest waters eventually return to the surface in the north Pacific. On this global journey the conveyor belt transports both energy in the form of heat, and immense quantities of organic and inorganic matter, and nutrients around the globe. Where these currents have nutrient-rich upwellings, such as the Benguela Current off the West Coast of Africa, fisheries are particularly productive. Another significant effect of THC is that the descending cold currents carry carbon dioxide into the deep waters away from the atmosphere.

The heat energy transported by the Gulf Stream is equivalent to the output of **1,000** *nuclear power stations*

Though scientists have long believed the Gulf Stream to be largely responsible for Europe's mild winters, recent ocean climate modelling suggests that simple atmospheric advection plays a larger role. In other words, Europe is warmed more by south-westerly winds which bring in warm maritime air, while eastern North America is colder because north-westerly winds freight in frigid continental air. There is mounting evidence that air currents, not ocean currents, account for the differences between the maritime climate of north-western Europe and the colder continental climate of eastern North America.

THERMOHALINE CIRCULATION

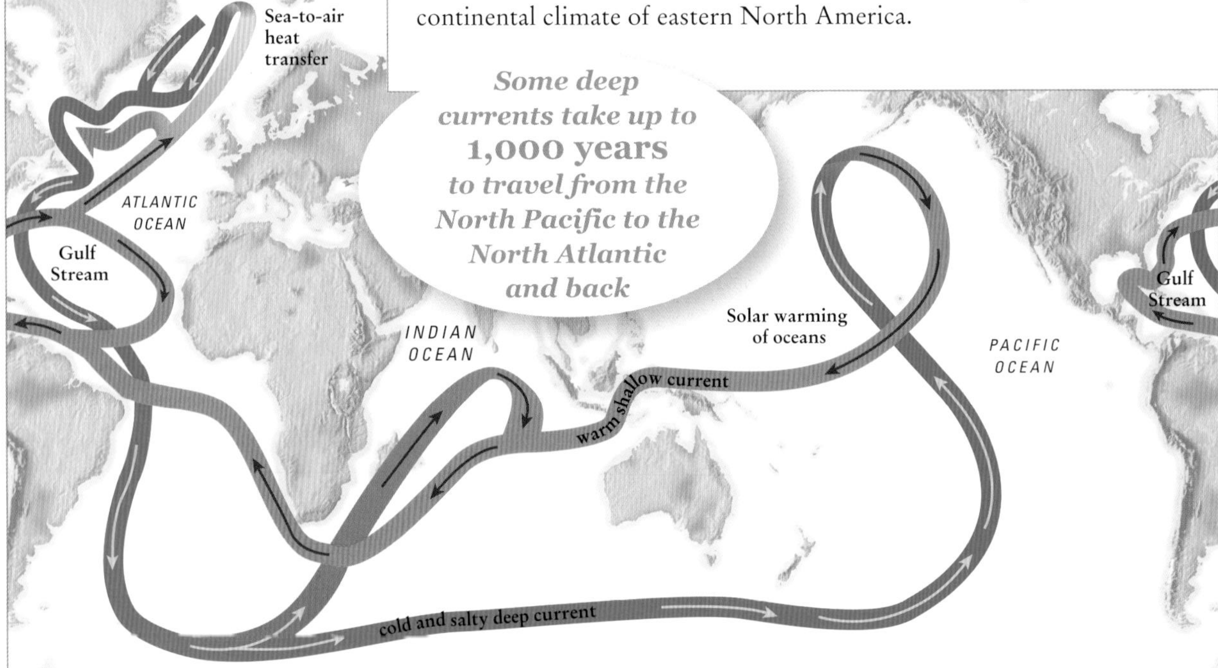

Some deep currents take up to **1,000 years** *to travel from the North Pacific to the North Atlantic and back*

The Gulf Stream

As can be seen on the map, the Gulf Stream is actually composed of three major currents that form one continuous system: the Gulf Stream proper, which turns into the North Atlantic Drift, then becomes the Norwegian Current when it reaches northern Europe. Along its route, the Gulf Stream and its companion currents release heat and moisture into the atmosphere. Though the Gulf Stream may exert some moderating influence over Europe's climate, recent research suggests that prevailing winds from the southwest have a larger climatic impact in creating an environment in which it is possible to grow palm trees in Cornwall in the UK, and crops near North Cape, areas that would otherwise be far too cold for either.

Upwellings

Deep currents which rise to the surface are often rich in nutrients, including nitrates and phosphates; these fecund ocean areas most often occur along the west coasts of the continents and are known as upwellings. Most of these areas exhibit high levels of primary production which contribute to the growth of phytoplankton – the basis of the marine food chain. Worldwide there are five major coastal currents associated with upwellings which support major fisheries: the Canary Current off Northwest Africa; the Benguela Current off the Atlantic coast of South Africa, Namibia and Angola; the California Current off the coasts of California and Oregon; the Humboldt Current off the coasts of Peru and Chile and the Somali Current or East Africa Coast Current.

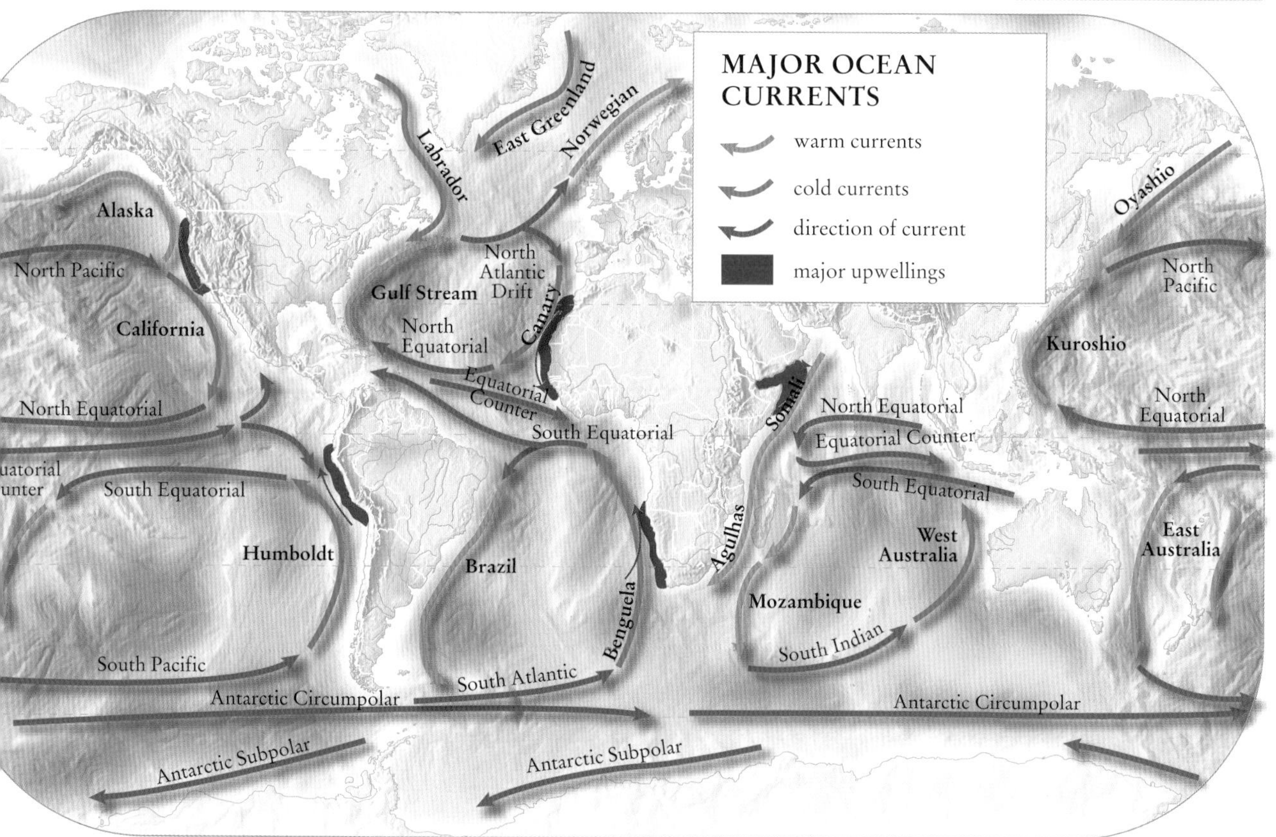

Rising Seas

As the world's climate changes, sea levels are rising inexorably. If current trends continue the world could face catastrophic coastal inundations creating millions of climate change refugees.

There is little doubt that the global climate is changing. According to the Intergovernmental Panel on Climate Change (IPCC), greenhouse gases, in particular carbon dioxide and methane are behind the rise in global temperatures.

As temperatures rise the climate is becoming more unstable, triggering more frequent and devastating cyclones, hurricanes, and storm surges, which inundate coastal areas, drowning low lying cities and towns, and displacing millions of people.

The warming trend is also causing the oceans to expand (warmer waters undergo thermal expansion, and take up more room, resulting in rising seas). During the twentieth century mean sea level rose by 15 to 20 centimeters (6 to 8 inches). If current emissions trends continue, the Intergovernmental Panel on Climate Change (IPCC) forecasts that by 2100 the seas could rise by between 20 and 90 centimeters (8 to 35 inches). A one meter (three feet) rise would inundate 15 percent of Egypt's arable land and 20 percent of Bangladesh's. The Maldives, an island chain off the southern coast of India, would disappear, as would a number of other low-lying islands in the Indian and Pacific Oceans.

BANGLADESH

heavily populated areas

area inundated by 1 meter sea level rise

A 1 meter sea level rise would affect 15 million people and submerge 17,000 km² of land.

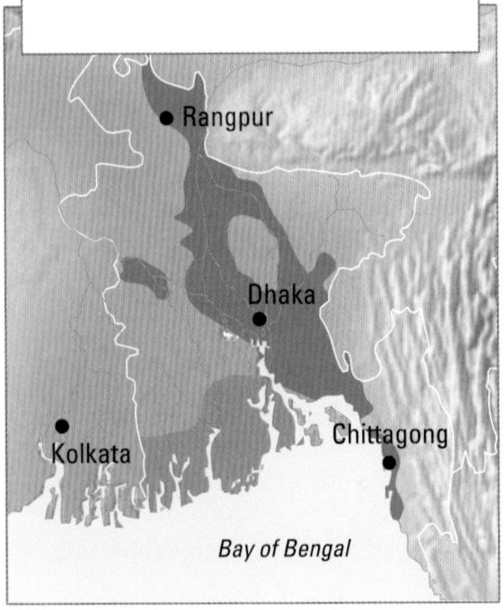

SEA LEVEL RISE

1870–2009

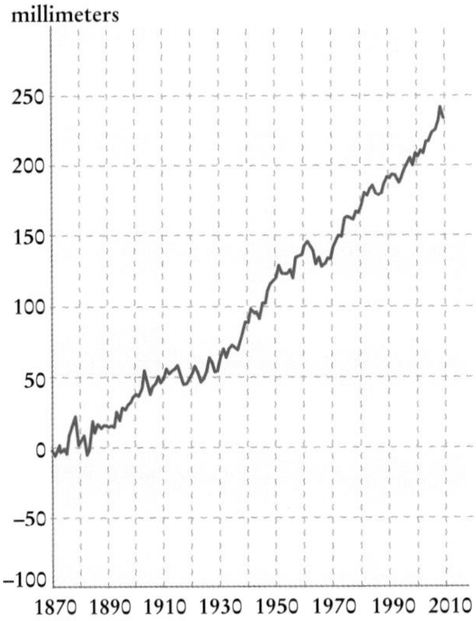

NILE DELTA

area inundated by a 1 meter sea level rise

A 1 meter sea level rise could affect 15% of Egypt's habitable land.

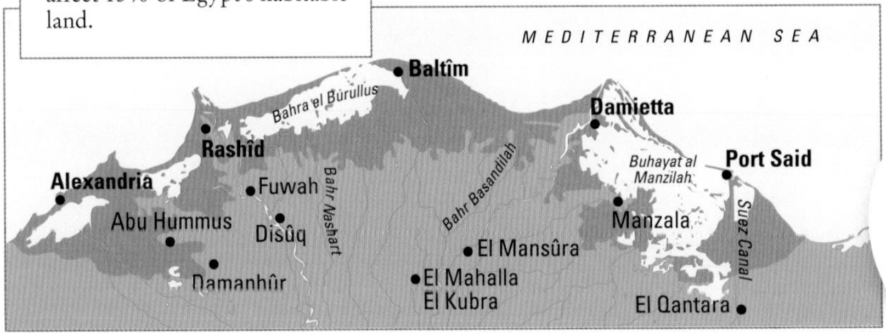

85% *of the Antarctic glaciers are in retreat*

An even more worrying scenario derives from new analysis of the geological sea level record, carried out by scientists at Princeton and Harvard universities, revealing that the planet's polar ice sheets are more vulnerable to large-scale melting under moderate global warming scenarios than previously thought. The melting of the Greenland and Antarctic ice sheets would add significantly to the ocean's volume, leading to a large and relatively rapid rise in global sea level, far exceeding any of the IPCC's forecasts. The study's authors suggest that the historical record supports a possible sea level rise of six to nine meters (20 to 30 feet) as a result of a 2°C temperature increase.

Such a catastrophic rise would inundate low-lying coastal areas where hundreds of millions of people now reside. It would permanently submerge New Orleans and much of southern Louisiana, along with most of southern Florida (including the Everglades) and other parts of the US East Coast, including New York city. Close to half of Bangladesh would be under water, along with most of the Netherlands, unless unprecedented and expensive coastal protection measures were taken.

EFFECT OF SEA LEVEL RISE

People at risk from 44 cm sea level rise (IPCC medium scenario A1B, assuming a 2.6°C temperature increase)
2100

- 13 million – 17 million
- 1 million – 8 million
- 100,000 – 999,999
- 10,000 – 99,999
- 1 – 9,999
- none or no data

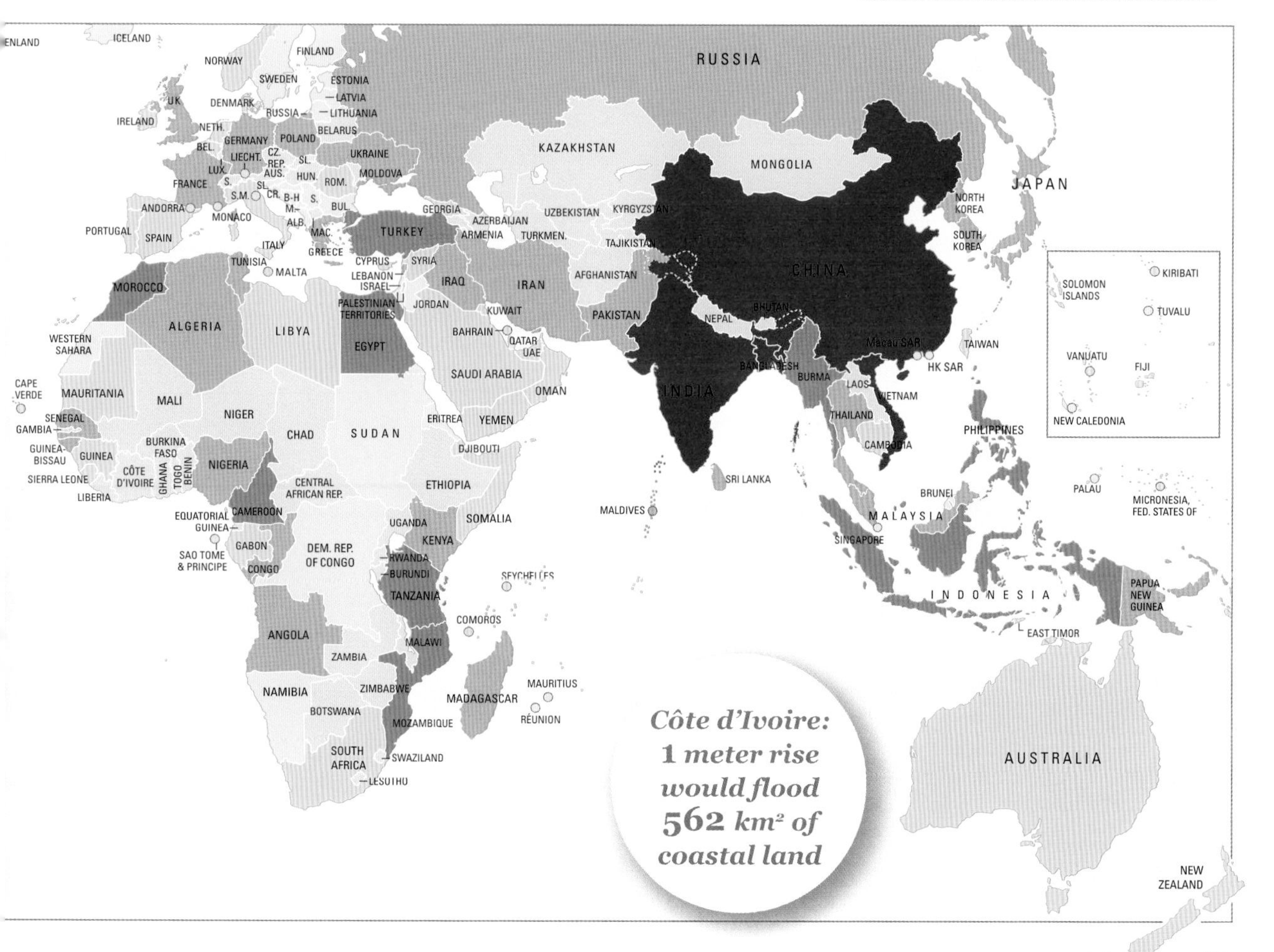

76–77, 78–79, 82–83, 84–85

Rising Seas: Small Island Developing States (SIDS)

The Small Island Developing States (SIDS) that are located in the Pacific and Indian Oceans include some of the smallest and most remote countries on earth. Although these countries are among the least responsible for climate change, due to a combination of physical characteristics, remoteness, and poor infrastructure they are likely to suffer most from its effects.

The IPCC forecasts a marked increase in the intensity and frequency of extreme weather events which, for many islands, will cause inundation, widespread damage to infrastructure, and loss of coastal agriculture. Small island nations are inherently vulnerable to natural disasters, including storm surges, floods, droughts, tsunamis, and cyclones. This vulnerability was highlighted by the 2005 Indian Ocean tsunami that devastated small islands in the Indian Ocean such as the Maldives and the Andaman and Nicobar islands, displaced more than 5 million people, and left more than 150,000 people dead.

30% *of known threatened plant species are endemic to these islands*

Sea level rises are expected to affect the low-lying atoll states of the Maldives, Tuvalu, Kiribati, the Marshall Islands, and Tokelau. Even where islands are not expected to face total inundation, there is a risk of increased waterlogged land, salinization of aquifers, and loss of fertile and residential areas.

The Carteret Islanders of Papua New Guinea may become the world's first climate change refugees around 2015, when their island home is expected to be over-run by rising seas. The typical sea level around the Carteret Islands has risen steadily over recent years, covering formerly inhabited areas, and salinating freshwater supplies and subsistence crops. Most of the islanders are expected to move to neighboring Bougainville, which will likely put them at odds with the island's long-term residents.

Maldives

Perhaps the country with the largest affected population is the Maldives, in the Indian Ocean, off the south coast of India. In 2010, 194 of the country's 1,192 island and coral atolls were inhabited by some 396,000 people. Since their average elevation is no more than 1.5 meters above sea level, and the highest point is only 2.4 meters above sea level, the entire archipelago is in imminent danger of disappearing beneath the sea. It is already suffering from coastal erosion, salt water intrusion, and damaging storm surges. The main island of Male, where half the population resides, will be completely submerged by a 1.5 meter rise, which could happen by 2100. Male's groundwater supplies are already severely depleted and have become increasingly saline. Almost the entire population is now served with desalinated water.

INCREASING RELIANCE ON WATER DESALINATION

Supply capacity in cubic meters per day
2002

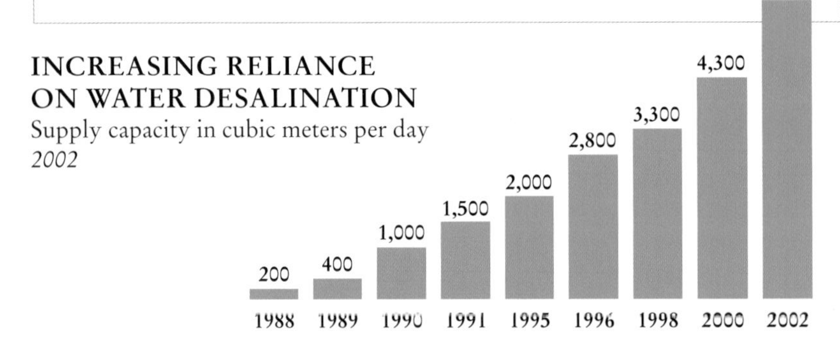

74–75

Tuvalu

Tuvalu is one of the island states assessed to be "extremely vulnerable" by the Environmental Vulnerability Index. Lying in the middle of the Pacific Ocean, its total land mass is 26 km² (10 square miles), its widest point is only 400 meters across and its highest point is only 3 meters above sea level. The islands that comprise the country are already experiencing repeated lowland flooding and coastal erosion, and are suffering from salt water intrusion, which has poisoned drinking water and depressed food production.

Realizing that the island's 11,000 residents cannot remain on the island much longer, the government appealed to Australia and New Zealand to accept their citizens. In 2001, the New Zealand government agreed to take in the entire population, if necessary. Currently, a quota of Tuvaluans is accepted each year.

VULNERABLE ISLANDS

Environmental Vulnerability Index assessment of SIDS

2005

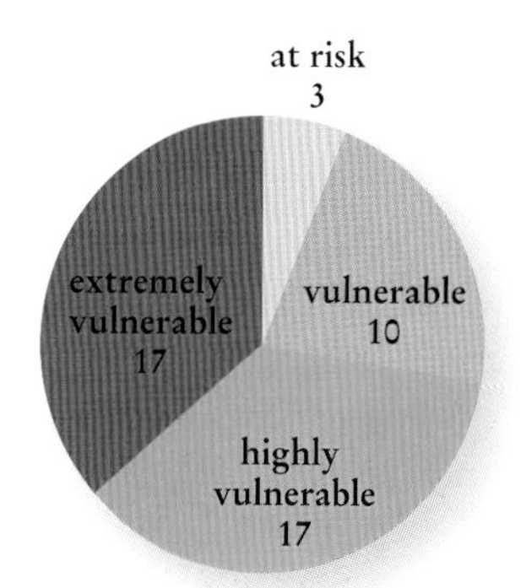

The 51 SIDS produce **less than 1%** *of the world's greenhouse gas emissions, but will be among the countries worst affected by climate change*

Kiribati

In 1999 two uninhabited islands, Tebua Tarawa and Abanuea, were inundated by rising seas.

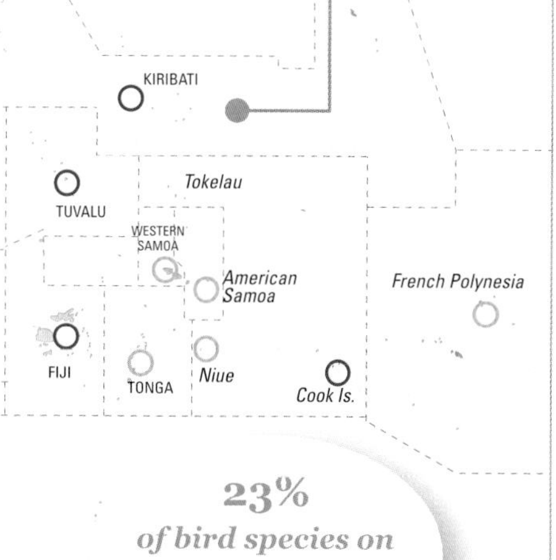

23% of bird species on these islands are threatened

VULNERABLE SMALL ISLAND DEVELOPING STATES (SIDS)

2010

- Pacific and Indian Ocean SIDS most at risk from sea level rise
- other Pacific and Indian Ocean SIDS

Extreme Weather Events

Over the past two decades, the world has experienced some of the worst and most destructive extreme weather events ever recorded; increasingly they are linked to climate change.

Severe weather in the context of oceans and coastal areas takes the form of hurricanes, which are also known as cyclones and typhoons, as well as floods, river-borne flooding, rogue waves, landslides or mudslides, dust storms, and heat waves. The International Disaster Database indicates that the strength, duration, and intensity of hurricanes and typhoons have increased by some 50 percent since the 1960s. While it is impossible to link individual weather events to global warming, the increase and intensity of events is in line with IPCC predictions of more extreme weather events being caused by climate change, and recent peer-reviewed studies which suggest a link between higher sea surface temperature and storm frequency. However, this is occurring at a lower level of global temperature rise than previously anticipated.

As the intensity of storms has increased, so has the human and financial cost. Hurricane Katrina, which savaged the Gulf Coasts of Louisiana, Mississippi, and Alabama in 2005, killed over 1,800 people, and caused property damage variously estimated at between $125 – $250 billion. However, such high profile developed world disasters can distract from the reality that developing countries are

El Niño

El Niño, also known as the Southern Oscillation or ENSO, is a periodic climate anomaly affecting the tropical and mid-Pacific Ocean on average every five years. During El Niño events warm water spreads across the Pacific from west to east, altering the weather patterns by bringing rain in the normally dry eastern Pacific and withering temperatures and drought in the western Pacific resulting in extreme weather such as floods, droughts and rising or cooling surface temperatures.

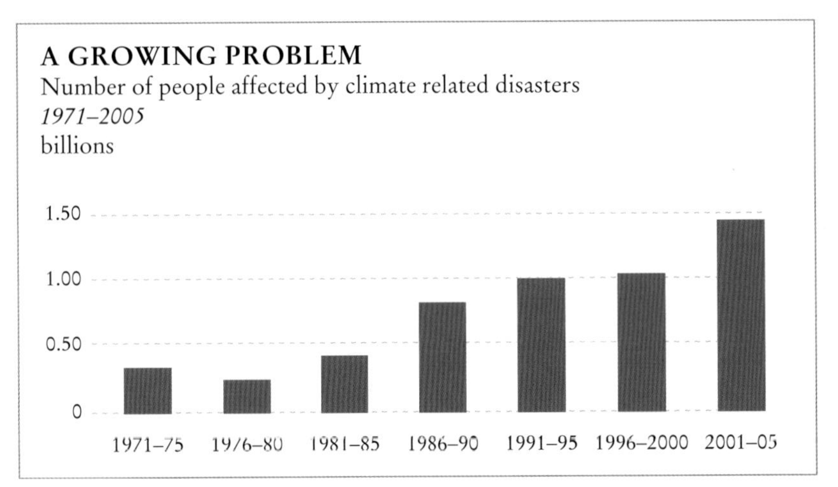

24–25, 26–27, 74–75, 76–77

those most affected by natural disasters, with a combination of poor infrastructure and populations concentrated in vulnerable areas such as along river banks and low lying coastal areas, wetlands, and swamps, adding to the effects. In 2008, Cyclone Nargis in Burma, for example, caused 85,000 fatalities and financial losses of $4 billion.

It has been suggested that the 2010 floods in Pakistan, in which half the average monsoon rains fell in one week rather than over the usual three month period are also the result of climate change affecting the pattern of the monsoon system. Flooding affected 14 to 17 million people and killed an estimated 1,700. The UN announced the scale of the disaster was greater than the effect of the Asian tsunami, and the Haiti and Kashmir earthquakes combined.

Unfortunately, it seems likely that those countries currently most affected by extreme weather events will continue to be the most impacted in the future.

WHO IS MOST AT RISK?

Climate Risk Index ranking of countries most affected by extreme weather events *1990–2008*

1–10, *most affected*
11 – 25
26 – 50
51 – 100
101 – 150
151 and above, *least affected*
no data

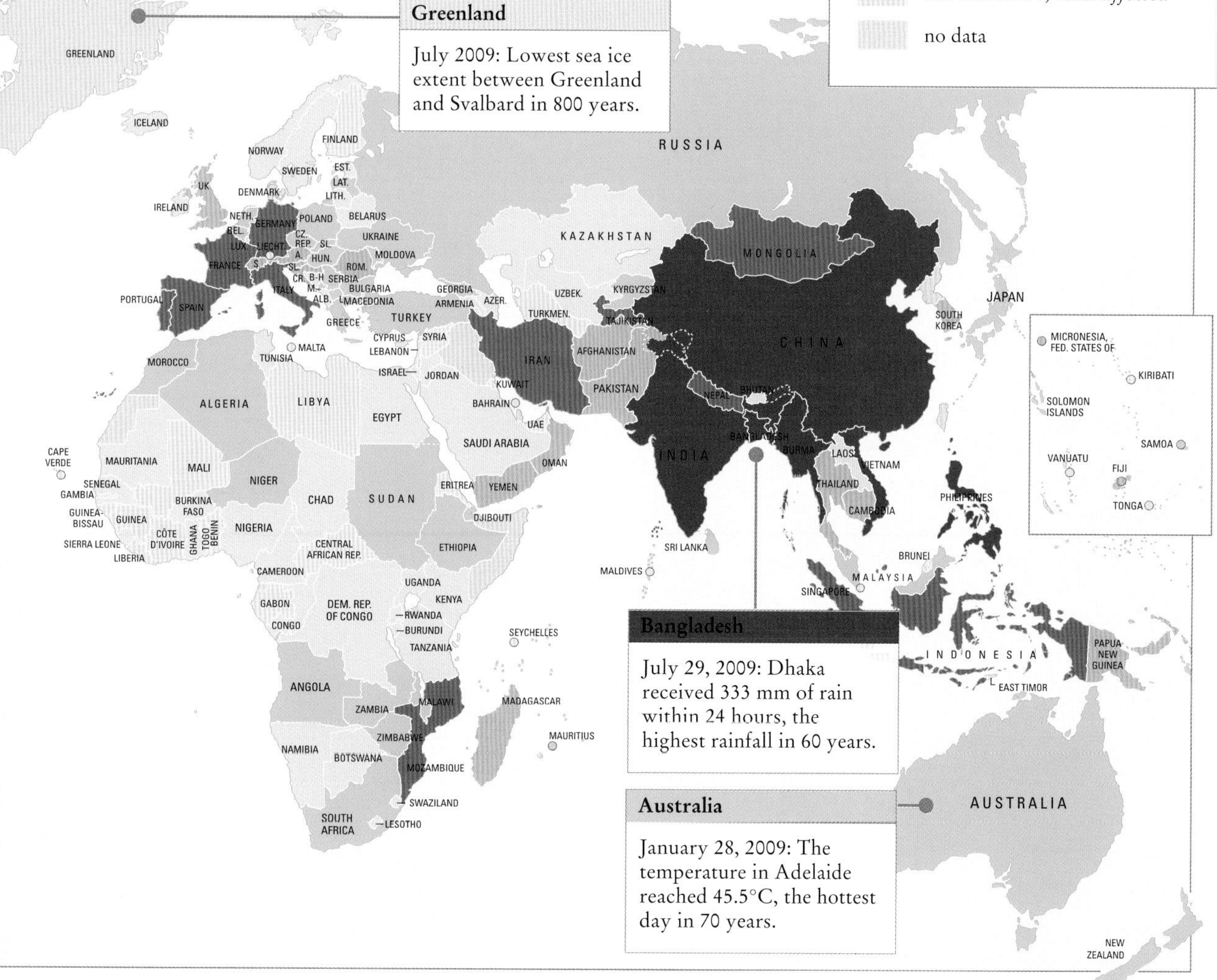

Ocean Acidification

If carbon dioxide emissions continue to rise so will the acidity of the ocean, interfering with the ability of creatures that manufacture calcium carbonate to build their shells or skeletons.

The oceans are a critical part of the planet's carbon cycle, recycling carbon dioxide between the atmosphere and the ocean. Some two billion tons of carbon have been captured this way over the past several decades. Excess carbon dioxide reacts with salt water to create carbonic acid. Over the past 250 years, the oceans have processed around 530 billion tons of CO_2, increasing ocean acidity by an average of 30 percent. Cold waters absorb more CO_2 than warmer ones, so the poles are particularly badly affected by acidification, showing a greater increase in acidity than warmer waters.

The chemistry of this process is well understood, but its consequences are not. Scientists fear that the impact of an increase in absorbed carbon dioxide, and therefore in acidity, could be far reaching and potentially catastrophic.

The organisms most affected by acidifying oceans are the "marine calcifiers", such as reef-building corals, molluscs, crustaceans, many species of plankton, and some species of algae. All of these calcifiers build skeletons or shells out of calcium carbonate, but an increase in carbonic acid in seawater dissolves the calcium carbonate, depriving some corals of their homes, and others of their protective shells or skeletons.

Since nearly one-third of all fish species live on coral reefs or are dependent on them for critical stages in their life cycles, the consequences for fisheries could be disastrous. Between 70 percent and 90 percent of all fish caught by coastal fishermen in tropical Asia are reef dependent at one time or another in their lifecycle, so acidification could have a significant impact on food security.

Marine scientists studying the potential impact of further ocean acidification are concerned that it could strike at the heart of the marine food chain by impeding the ability of *foraminifera* plankton to form their shells. *Foraminifera* plankton are very abundant in the oceans and are responsible for much of the carbon removal. Studies in the Southern Ocean indicate that the current shell weights of *foraminifera* are 30 to 35 percent lower than the weights of shells that have been found in sea sediments that are thousands of years old.

Scientists prefer to err on the side of caution, however, once the ocean's pH balance has been upset, it will take thousands of years to reverse it.

The ocean's recycling system upset

The ocean is an efficient recycling system. Through tidal fluctuations and currents, hundreds of species on the seabed clean the water and recycle nutrients making them available for plankton – the basis of the ocean food chain. One seabed dwelling organism that assists this fundamental process is the brittle star. Acidification decreases the star's ability to make its calcium carbonate skeleton, thus impeding its ability to capture nutrients.

PAST AND PROJECTED OCEAN pH LEVELS
Future levels based on IPCC scenarios

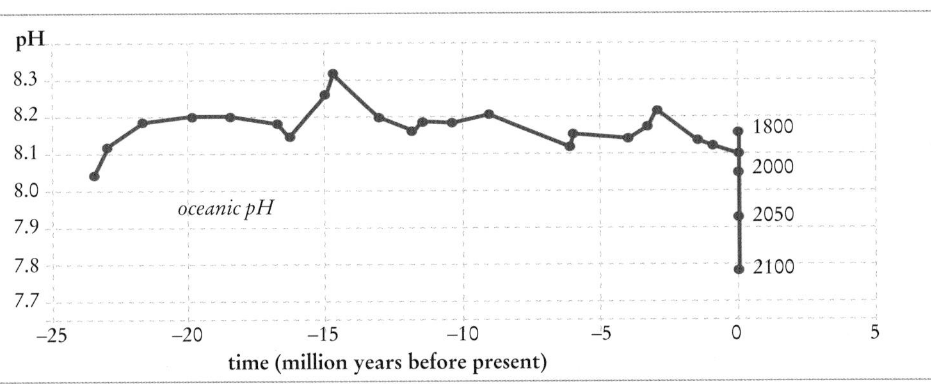

34–35, 48–49

Estimated change in annual mean sea surface pH between 1750 and 1994, calculated from measurements of total CO_2 and total alkalinity. The lower pH values represent an increase in acidity (from 0.05 to 0.11), as shown in orange and dark orange on the map. Colder water environments, particularly at the poles, are most affected.

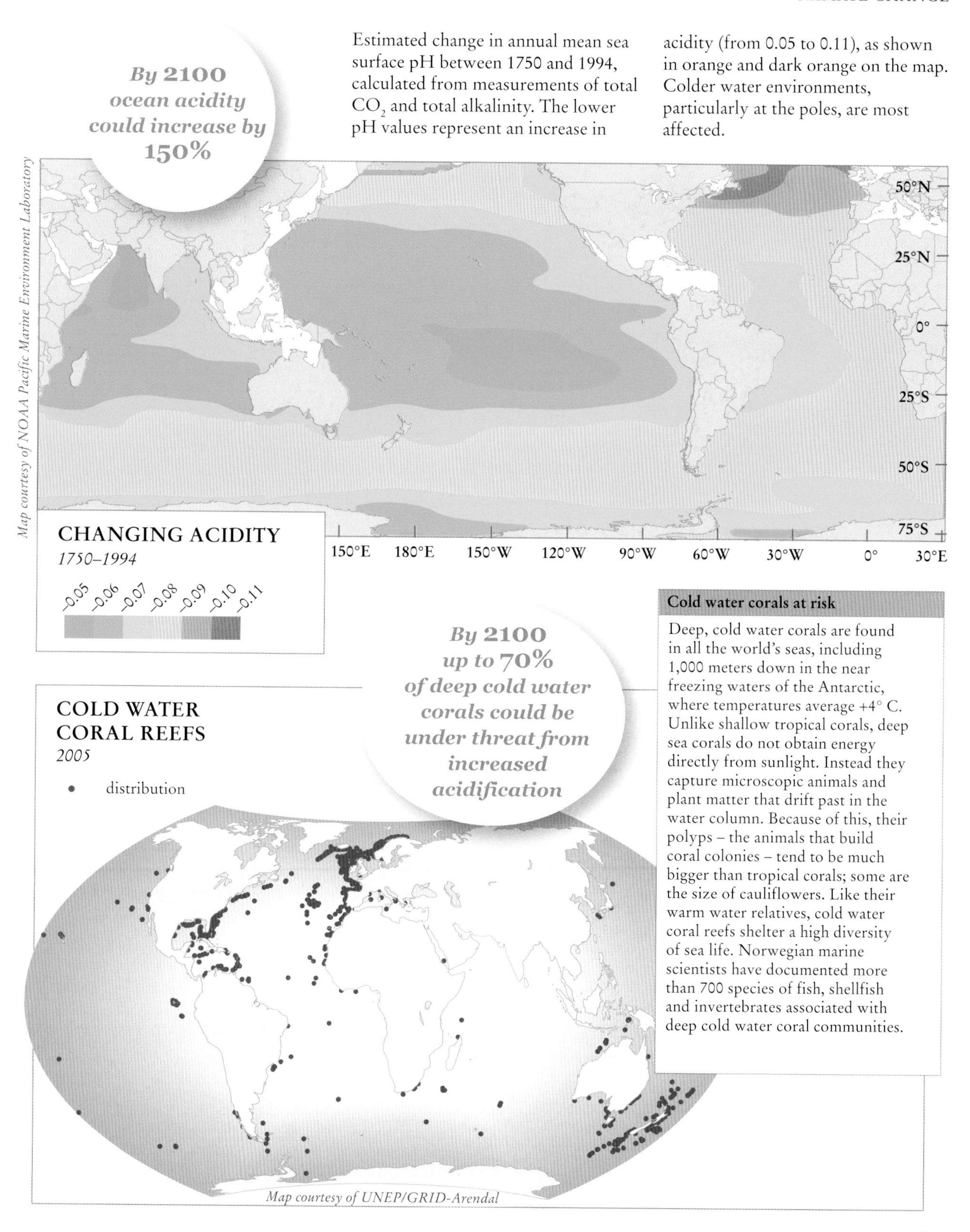

Cold water corals at risk

Deep, cold water corals are found in all the world's seas, including 1,000 meters down in the near freezing waters of the Antarctic, where temperatures average +4° C. Unlike shallow tropical corals, deep sea corals do not obtain energy directly from sunlight. Instead they capture microscopic animals and plant matter that drift past in the water column. Because of this, their polyps – the animals that build coral colonies – tend to be much bigger than tropical corals; some are the size of cauliflowers. Like their warm water relatives, cold water coral reefs shelter a high diversity of sea life. Norwegian marine scientists have documented more than 700 species of fish, shellfish and invertebrates associated with deep cold water coral communities.

Disappearing Arctic

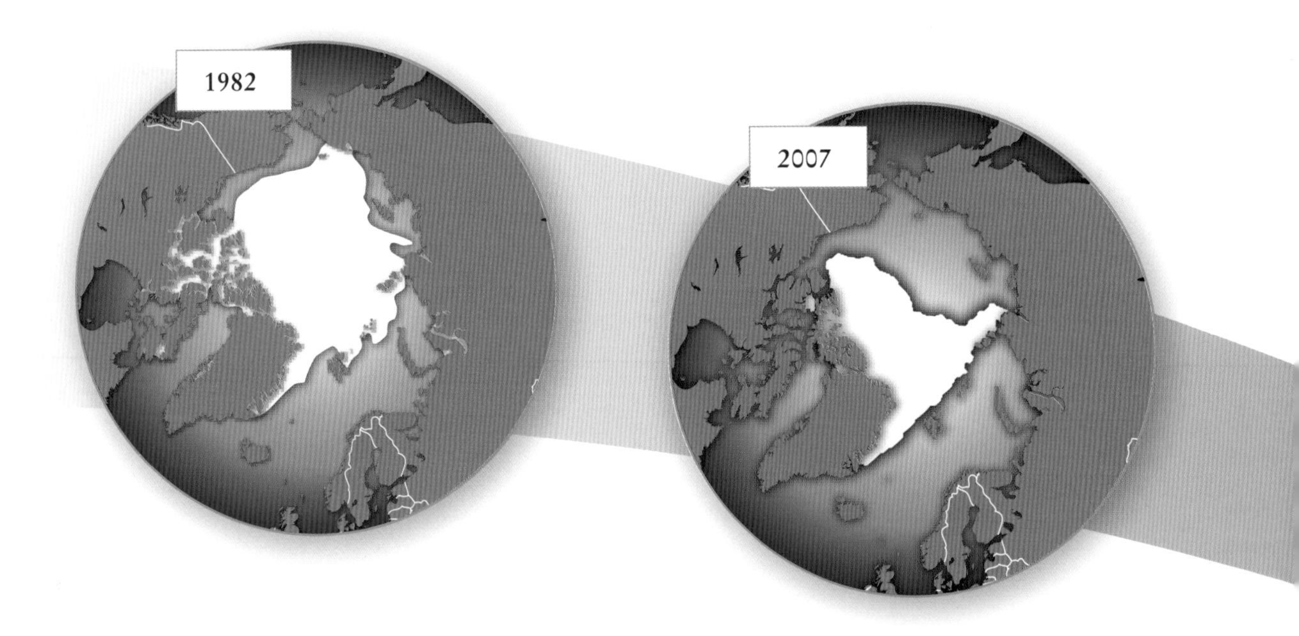

The extent of Arctic sea ice is shrinking as the polar temperature increases.

Satellite images of the Arctic confirm that, as global temperatures rise due to climate change, the Arctic sea ice is contracting. By 2007, the extent of arctic sea ice was nearly 40 percent below the minimum average for the period 1979–2000.

The poles are on the front lines of climate change, with potentially catastrophic impacts projected by the end of this century. Currently, the Arctic functions like a huge air conditioner for the northern hemisphere. Sea ice cools air and water masses, playing a key role in the circulation of ocean currents, and also reflecting solar radiation back into space. The loss of these functions could add to climate changes that are already taking place, potentially increasing the melt rate.

Recent studies of the last interglacial period, about 125,000 years ago, found that the polar ice sheets are subject to large-scale melting when average temperatures in the Arctic and Antarctic are between 3°C and 5°C warmer than at present. During this earlier geologic warming period both the Greenland and West Antarctic ice sheets shrank, considerably and sea levels were 6 to 9 meters (20 to 30 feet) higher than today.

Already, scientists have found that the discharge rate (melt water runoff) from the Greenland ice sheet is 30 percent higher compared to rates during the last decade. Over the last six years Greenland has lost 183 gigatons of ice per year, or the equivalent of 200 square kilometers (77 square miles).

Not only is the temperature increasing and melting continental ice caps, the sea ice in the Arctic is not forming as rapidly nor as thickly as in previous decades. In January 2010, Arctic sea ice grew by only 34,000 square kilometers (13,000 square miles) a day, a rate roughly one-third of the pace during the 1980s. Since it is thinner, it also melts more quickly during the summer months.

The environmental impacts are already being felt. Thawing permafrost in the Arctic has damaged houses, roads, airports, and oil and gas pipelines,

Impact on Arctic wildlife

Arctic wildlife populations are being adversely affected by melting sea ice, coastal erosion, and disruptions in availability of seasonal food sources. Polar bears in particular have experienced a precipitous decline in birth rates as adults need sufficient sea ice from which to hunt for seals and other prey. Retreating sea ice has reduced the number of resting platforms needed by walruses while they search for food, leading to unusual migrations to land – in 2010, 10,000 to 20,000 walruses congregated along the shore of Alaska. The US Geological Survey predicts a 40 percent decline in walrus numbers by 2095.

 ◀◀ 74–75, 76–77

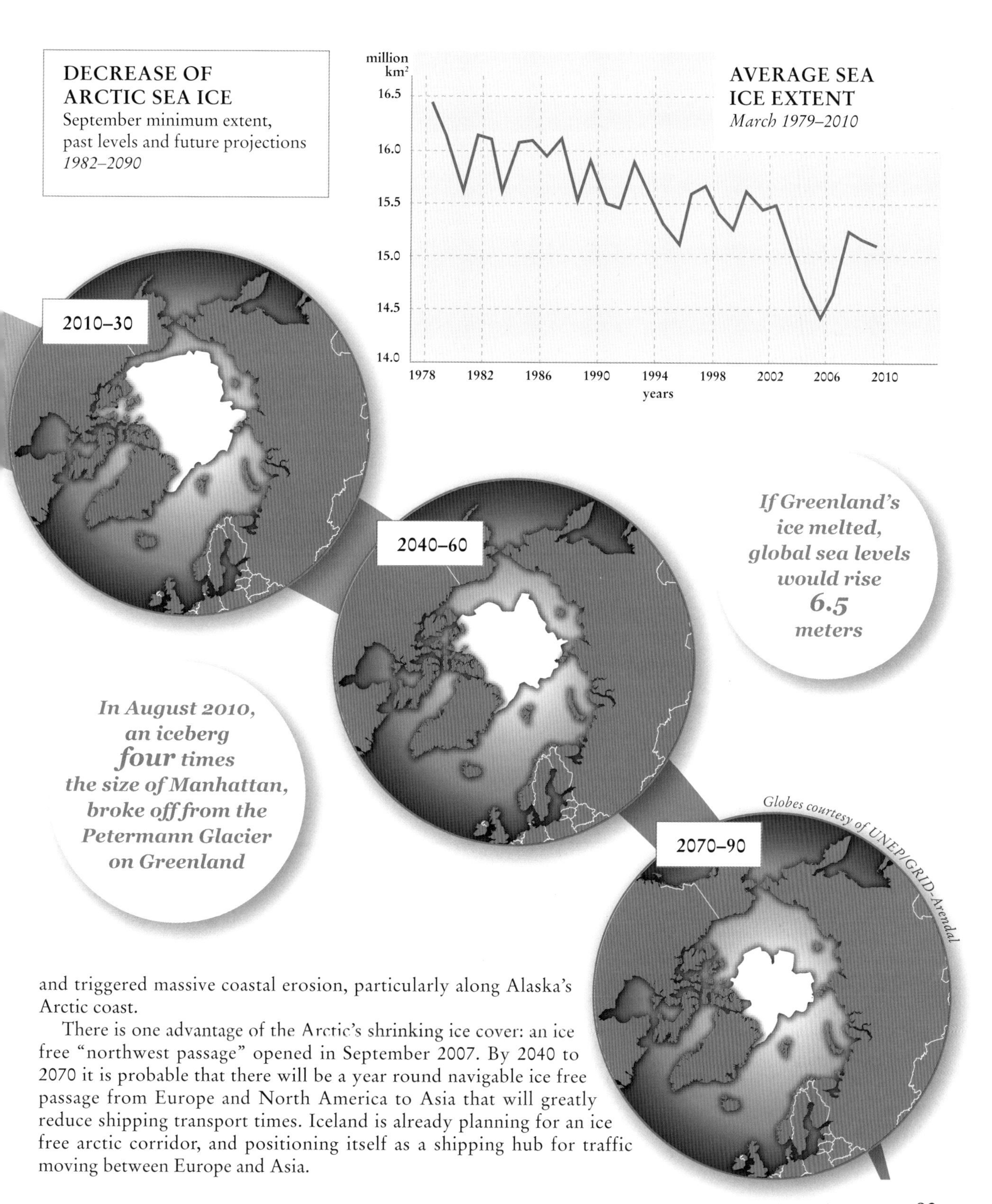

and triggered massive coastal erosion, particularly along Alaska's Arctic coast.

There is one advantage of the Arctic's shrinking ice cover: an ice free "northwest passage" opened in September 2007. By 2040 to 2070 it is probable that there will be a year round navigable ice free passage from Europe and North America to Asia that will greatly reduce shipping transport times. Iceland is already planning for an ice free arctic corridor, and positioning itself as a shipping hub for traffic moving between Europe and Asia.

84–85 ▶▶

Antarctic: Losing Ice Cover

Previously thought to be immune to the degradations of climate change, the West Antarctic ice sheet is now losing volume and a number of floating ice shelves on the Antarctic Peninsula have already broken apart.

The Antarctic is not immune to climate change. The frozen continent is divided into two massive ice sheets: the East Antarctic and the smaller West Antarctic. The Antarctic Peninsula is the strip of ragged land that points towards the southern tip of South America. This region has been hardest hit, warming by 3°C in the past 50 years, the most rapid rise seen anywhere in the Southern Hemisphere. As a result, although the massive East Antarctic ice sheet, which covers most of this frozen continent, is intact and stable, the smaller West Antarctic ice sheet has lost as much as 150 cubic kilometers (36 cubic miles) of ice per year over the past decade. While the Arctic is a sea covered by ice, the Antarctic is a land mass covered by ice. As the ice melts it will flow into the seas, adding greatly to its volume and causing sea levels to rise.

If the entire West Antarctic ice sheet – an area the size of Texas – were to melt it would raise global sea levels by between 3.5 and 6 meters (11 to 20 feet). Though the likelihood of this occurring is considered remote at this time, scientists are alarmed at the rapidity of the changes in ice mass.

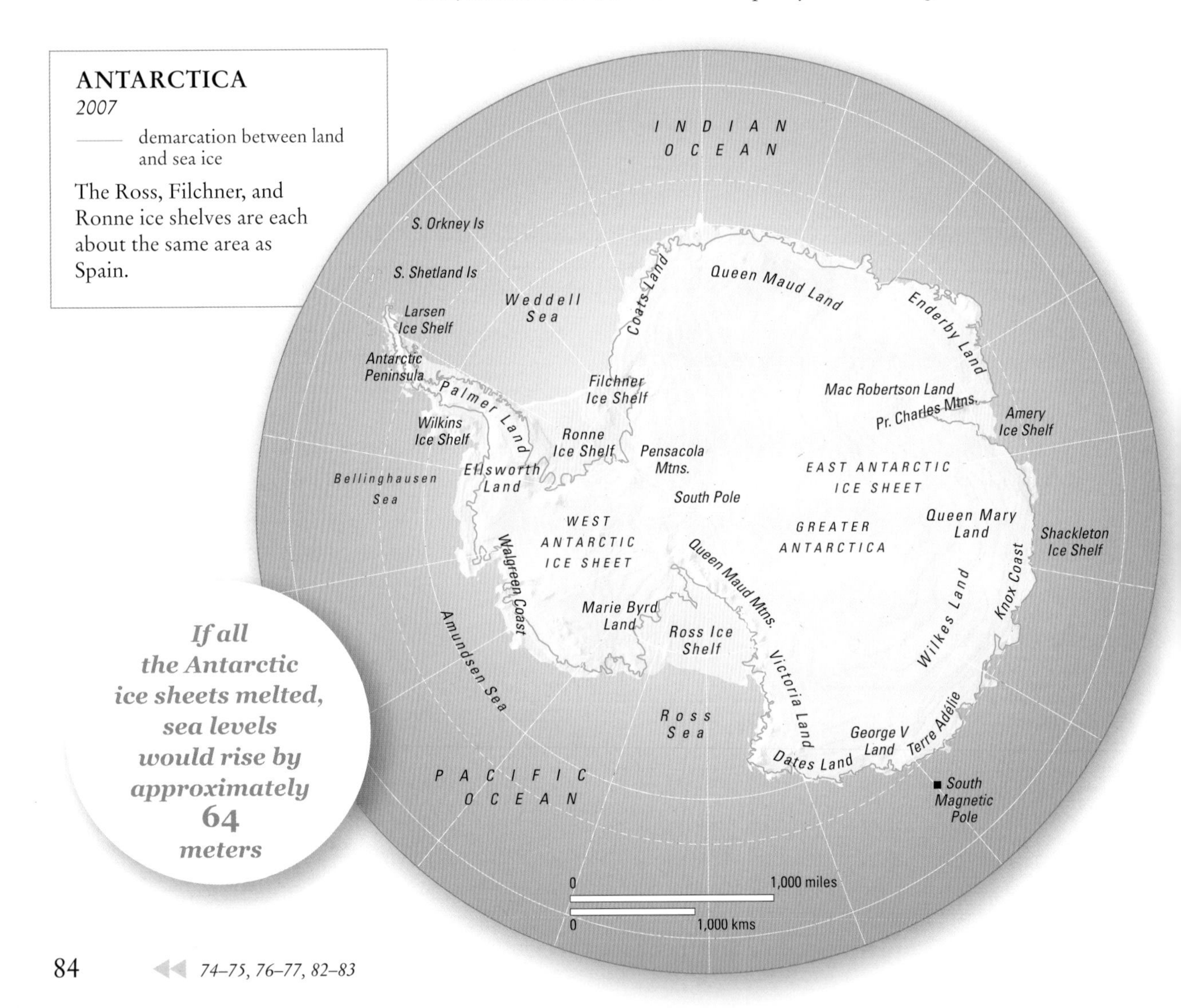

74–75, 76–77, 82–83

Another warning sign is the continued break-up of the massive West Antarctic ice shelves that rest on the ocean. Their demise does not raise sea levels, but the disintegration of floating ice shelves could portend more catastrophic changes in the future.

Over the past 50 years, scientists have documented the loss of around 25,000 square kilometers (9,653 square miles) of ice from ten floating ice shelves in the West Antarctic. In 2002, the massive Larsen B ice shelf on the Antarctic Peninsula, covering 3,250 square kilometers (1,255 square miles), broke apart and disappeared within a month. Climate change is chipping away at the vulnerable edges of the Antarctic.

There is some concern among scientists that the continued massive melting of both ice shelves and the West Antarctic ice sheet could alter the dynamics of the Southern Ocean's food web, disrupting the production of krill, one of the main grazers of phytoplankton and in turn a major source of food for many other species. The loss of krill is already reverberating up the food chain, affecting whales, seals, albatrosses, penguins, and carnivorous fish.

LOSS OF ICE MASS

Increasing rate of loss of West Antarctic ice mass

1974–2010

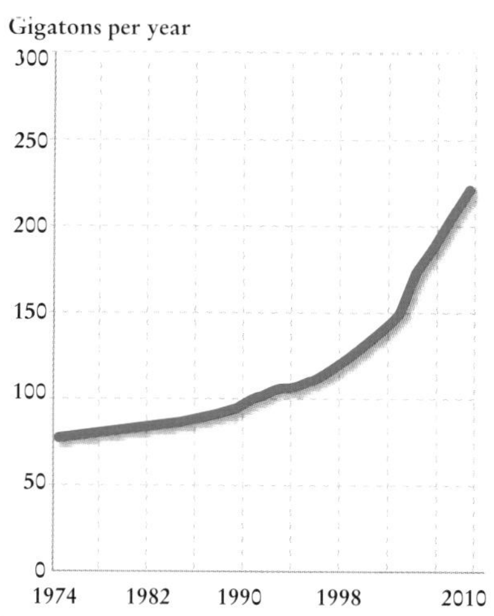

The Antarctic contains **90%** *of all the ice on earth*

Wilkins Ice Shelf

In 2008, the Wilkins Ice Shelf broke apart, taking with it some 500 km² (193 square miles) of floating ice.

March 6, 2008

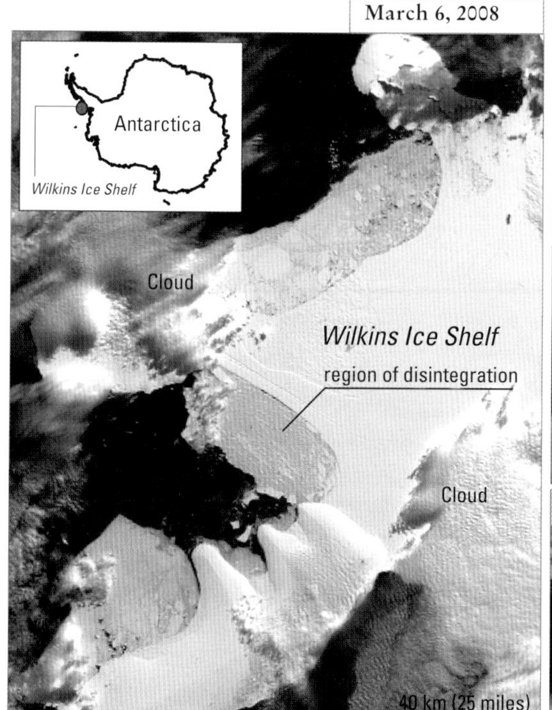

February 28, 2008

February 29, 2008

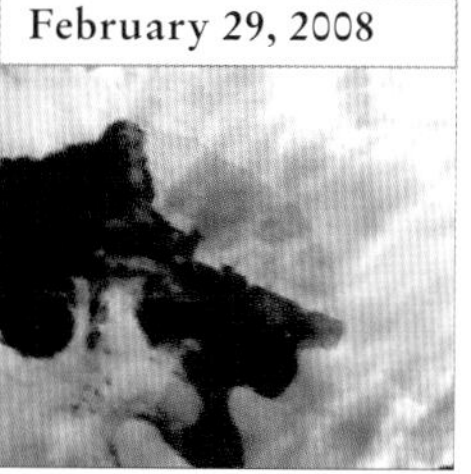

March 8, 2008

Adelie Penguins

Adelie Penguins depend on sea ice and a good supply of krill. As temperatures have risen both have declined. Researchers studying Adelie colonies midway down the Antarctic Peninsula have documented corresponding population declines of 80 percent since the 1970s, from 40,000 breeding pairs to 2,400 by 2008.

Nations extend territorial and asset conflicts to the sea.

Part Five: Seas in Conflict

Contested Islands

There are contested islands throughout the world's oceans, but four island groups and shoals in the South China Sea pose the greatest risk of igniting regional conflict between countries with competing and overlapping claims.

There are a number of on-going disputes over seemingly insignificant island territories, many of them dating back over decades (in one case, a century). With world hydrocarbon supplies on land fully exploited for the most part, the search for offshore reserves has accelerated, adding an ominous geo-political dimension to existing disputes where contested island areas hold rich deposits of oil, gas, and minerals.

The most hotly contested region is the South China Sea, a densely populated area encompassing 3.5 million square kilometers (1.4 million square miles) bordering on China, Taiwan, the Philippines, Vietnam, Malaysia, Singapore, Indonesia, and Brunei.

It is one of the most strategic bodies of water in the world, the primary shipping lifeline for Japan, China, South Korea, Vietnam, the Philippines, and Indonesia. More than half the world's merchant fleets pass through the Strait of Malacca, Lombok, and Sunda. Oil tankers transit over 9.5 billion barrels of oil per year through the Strait of Malacca alone.

Four island groups are at the heart of this on-going struggle – the Paracel Islands, the Scarborough Shoal, the Macclesfield Bank, and the Spratly Islands. The Paracel Islands consist of some 30 uninhabited reefs, islets, and sandbanks – surrounded by rich oil deposits. In 1974, Vietnam lost a naval

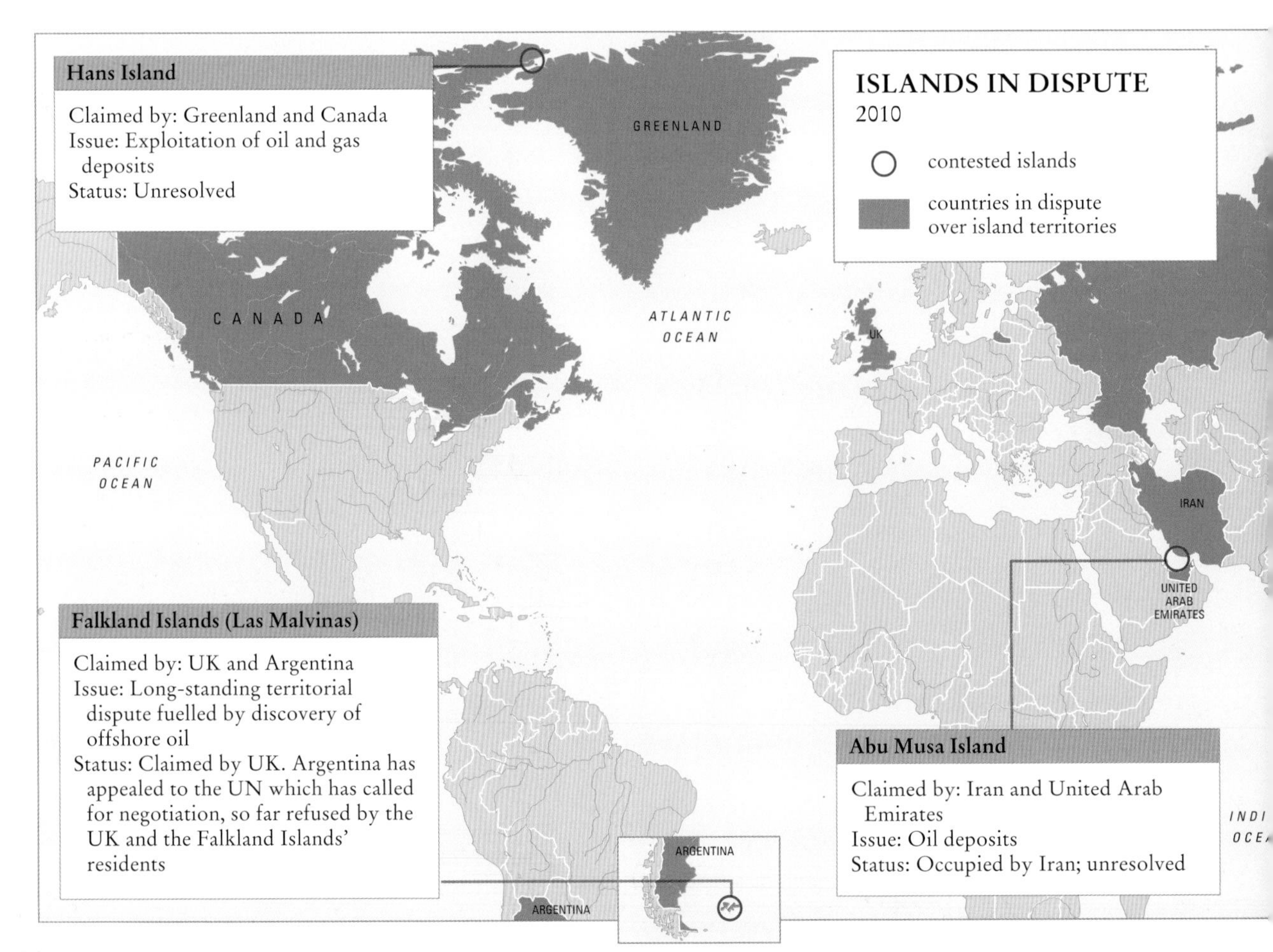

56–57

engagement with China over the Islands. China has occupied them ever since, rejecting claims from Vietnam and Taiwan.

The Spratly Islands, consisting of some 100 reefs, islets, atolls, cays, and islands covering some 400,000 square kilometers (154,000 square miles), are caught in the middle of a regional dispute involving six claimants. Of these, only Brunei has no military presence on the islands. China occupies eight islands, Taiwan one, the Philippines nine, Malaysia nine and Vietnam 27. In 1988, China sank three Vietnamese vessels in a dispute over Fiery Cross Reef in the Spratlys.

The islands are surrounded by very productive fisheries, and oil and gas deposits. Despite the fact that all parties involved in these disputes signed the Declaration on the Conduct of Parties in the South China Sea in 2002, reaffirming their commitment to freedom of navigation and over-flight, and pledging to resolve territorial disputes peacefully, the region is a hotbed of unresolved rivalries and competing claims.

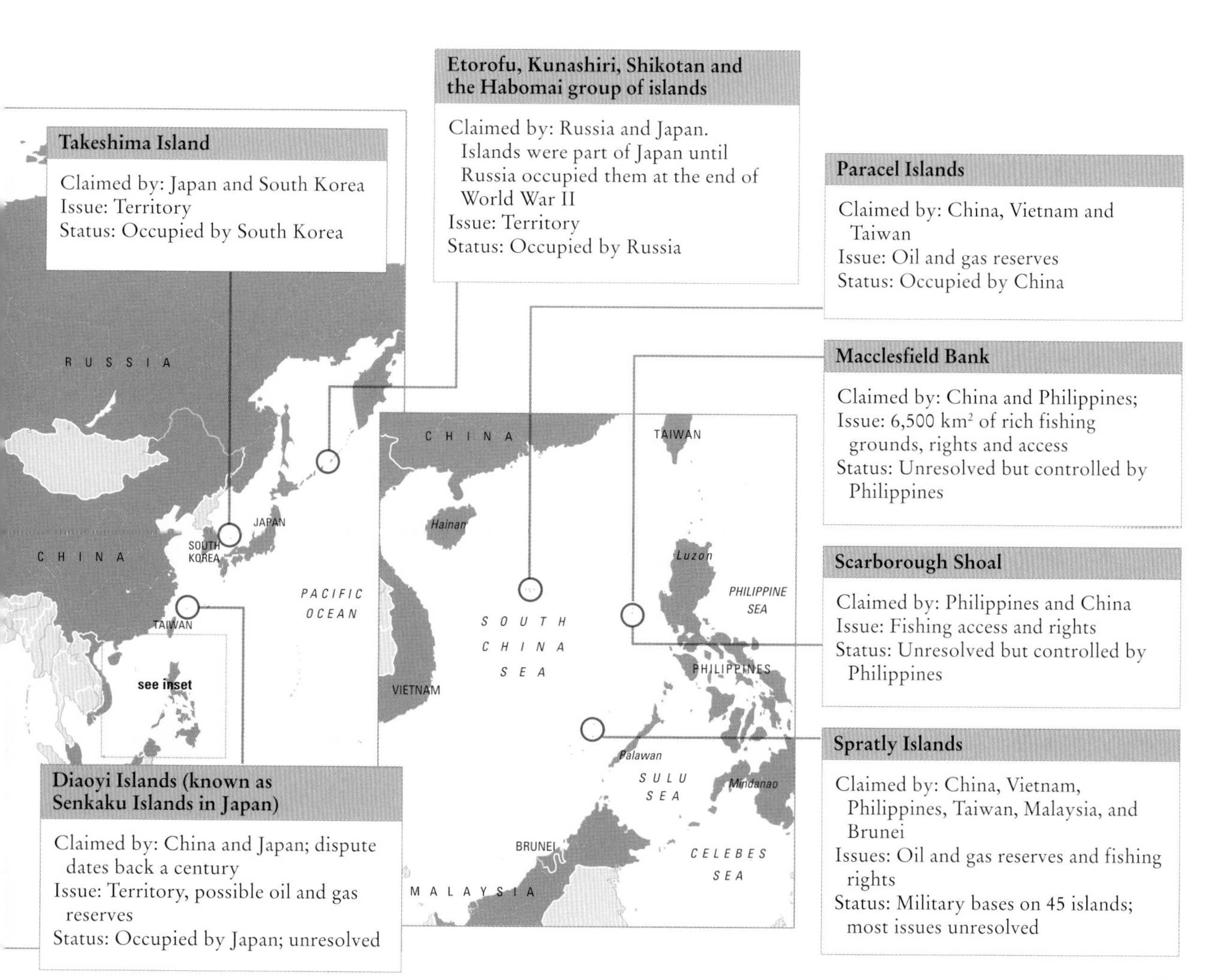

Piracy: A Recent Growth Industry

Global piracy is on the increase, putting shipping in strategic areas such as the Horn of Africa in jeopardy. Though global in nature, piracy is now concentrated in the Gulf of Aden and off the Indian Ocean coast of Somalia.

Pirates have been boarding and robbing ships for millennia. The earliest accounts of sea-going robbers date back to 1400 BC, when the Tyrrhenians preyed on shipping in the eastern Mediterranean.

Though eliminated from much of the world by 1900, there has been a marked resurgence of the practise during the past five years. In 2009, the International Maritime Bureau of the International Chamber of Commerce reported 406 incidents of piracy globally, a doubling of attacks over 2008. Of these, more than half – 217 incidents – were off the coasts of Somalia. Global piracy costs the maritime industry between $1 and $16 billion a year in extra insurance costs, negotiating and paying hijacking and ransom fees, and in implementing mitigation efforts.

Acts of piracy have increased in number every year since the middle of the last decade, due mainly to intensified piracy off the coasts of Somalia. Other incidents of piracy were scattered around the world, but concentrated in two regions: in the Gulf of Guinea, off the coast of Nigeria; and in South-East Asia, mostly off the coast of the large Indonesian island of Sumatra and in the Strait of Malacca. Other clusters of piracy occur in the South China Sea, in South Asia (mostly in or around the Bangladeshi port of Chittagong), and off both coasts of South America. However, for the most part, acts of piracy in areas other than Somalia are aimed at stealing goods, not in hijacking entire ships and taking hostages.

Between 2003 and 2008, there were 1,845 actual or attempted acts of piracy around the world

Piracy has actually fallen in South-East Asia, mainly as a result of concerted efforts at deterrence from the nations that surround the Strait of Malacca and nearby waters: Singapore, Malaysia, and Indonesia signed a cooperative agreement to fight piracy in September 2007, under the UN Convention on the Law of the Sea (UNCLOS).

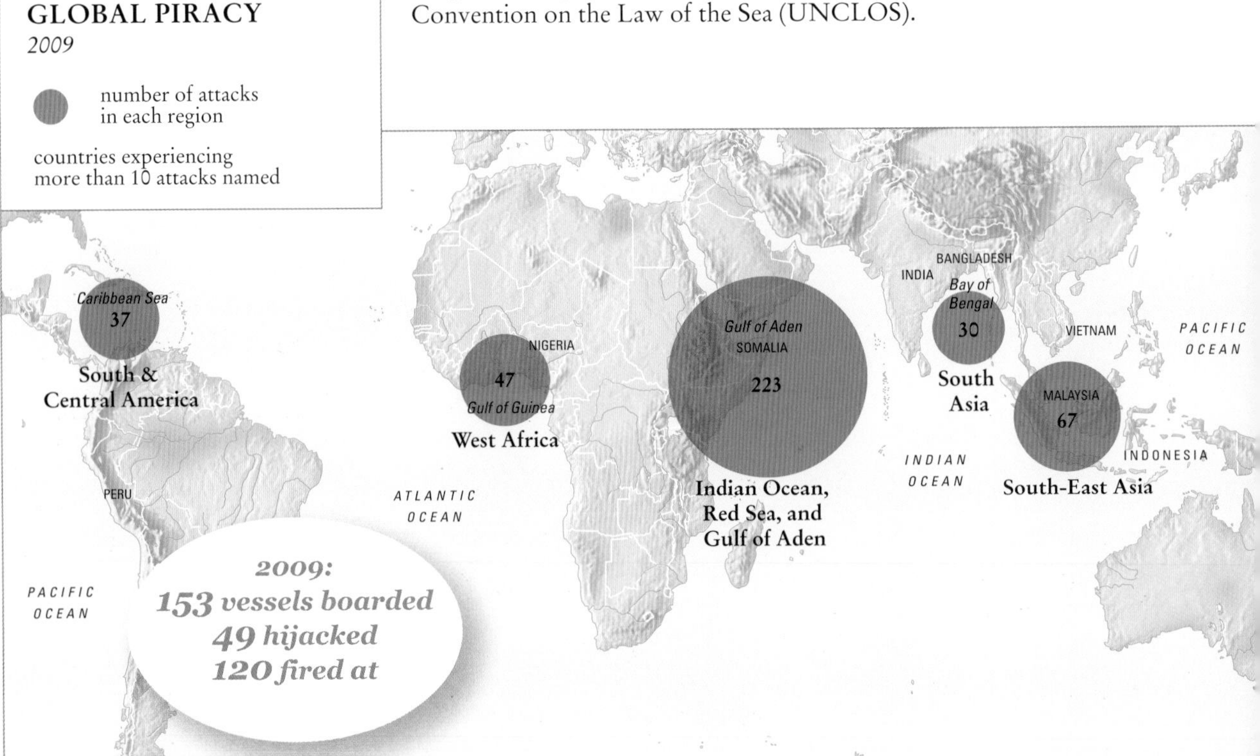

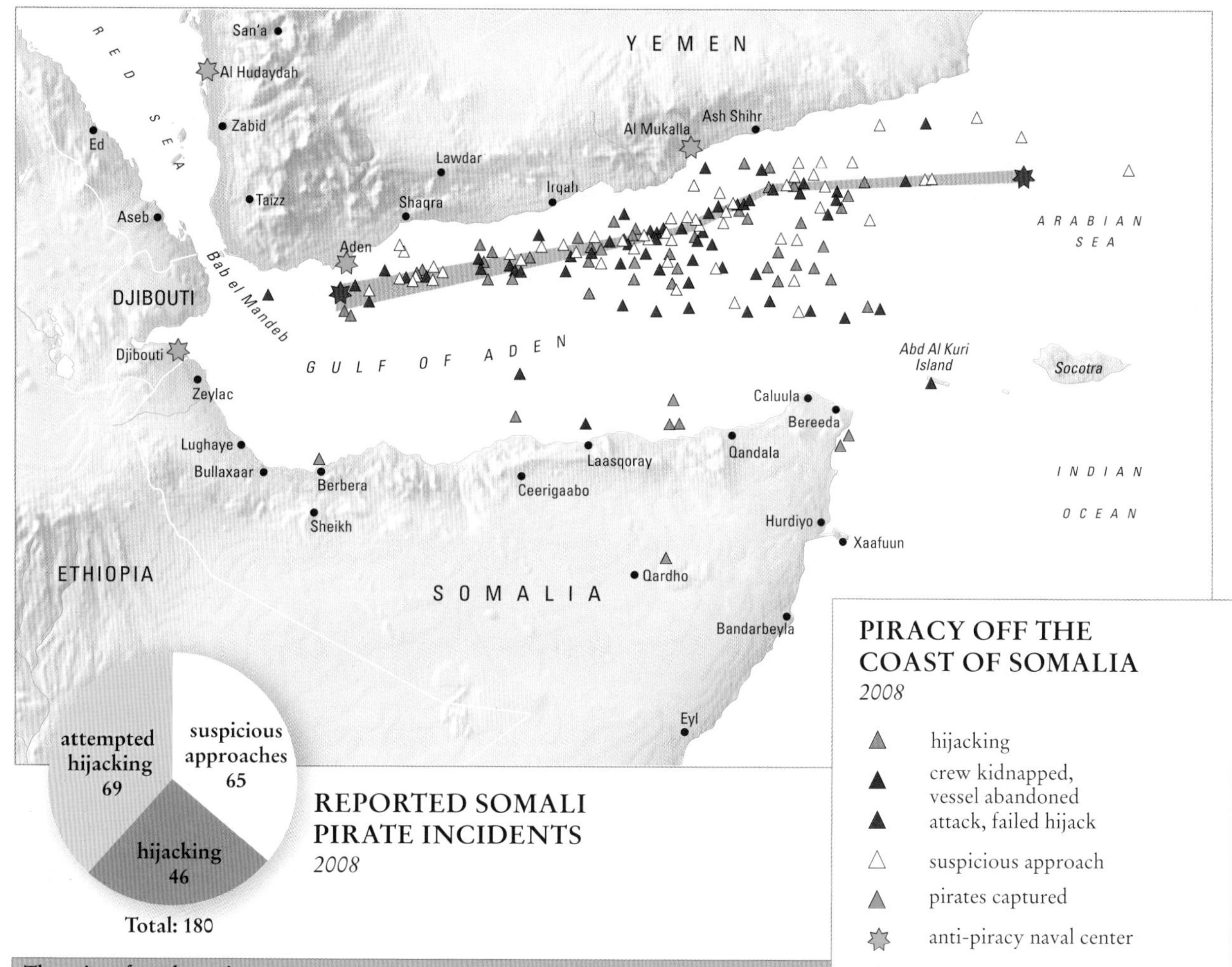

PIRACY OFF THE COAST OF SOMALIA
2008

REPORTED SOMALI PIRATE INCIDENTS
2008

The price of modern piracy

The Gulf of Arabia and the western Indian Ocean are the only regions in the world where the vast majority of attacks (perhaps all of them) are made with the intention of seizing the vessel in order to hold it, and the crew, for ransom. Virtually all successful hijackings were attributed to Somali pirates (47 incidents in 2009), accounting for 80 percent of crew members taken hostage.

Beset with civil war since 1991, Somalia has no effective government. As a result, it has become a haven for pirates; some 1,400 of them are based mainly along the coast in Central Somalia in Haradheere, and in the north-eastern region of Puntland in the Eyl district.

A worrying trend is the increase in attacks off the country's eastern and southern coasts, hundreds of kilometers offshore, in waters previously considered safe. In April 2010, Somali pirates seized three Thai fishing vessels approximately 1,900 kilometers (1,180 miles) off the coast of Somalia.

Somali pirates reportedly reaped $60 million from hijackings of ships and their crews in 2009, an increase on the estimated $30 to $50 million paid out in 2008. Given the state of the country, Somali piracy will likely continue to be a productive enterprise for the foreseeable future.

The presence of a multi-country naval fleet aimed at stemming piracy around Somalia is having some success.

An international flotilla consisting of war ships from Russia, France, UK, USA, China, Canada, Netherlands, Spain, Portugal, Norway, Denmark, Greece, Turkey, Italy, Thailand, and India were engaged in anti-piracy patrols from 2009 to 2010, responsible for covering some 6.4 million square kilometers (2.5 million square miles) of ocean.

Sharks are among the species that Large Marine Ecosystems are intended to protect.

Part Six: Management of Coastal and Marine Areas

INTEGRATED COASTAL & OCEAN MANAGEMENT

As of 2002, 145 of the 187 nations, territories and semi-sovereign states with a marine coastline, had launched integrated coastal management programs. The overwhelming majority of these initiatives involved management of coastal areas and resources at the sub-national or country regional level.

Since the mid-1970s, coastal management efforts have accelerated, with rapid progress over the course of the 1990s. By 2002 there were 622 different programs in 145 coastal countries and semi-sovereign states.

Integrated coastal and ocean management efforts date from 1965, when the San Francisco Bay Conservation and Development Commission was established. The Commission had total jurisdiction over San Francisco Bay, including the area up to 100 feet inland from the mean high-tide mark.

In 1972, the USA passed the national Coastal Zone Management Act, administered by the National Oceanic and Atmospheric Administration (NOAA). Its primary focus was to provide the nation's 35 coastal states and territories with guidance and funds for program planning and implementation.

Debate continues as to what constitutes a credible integrated coastal and/or ocean management plan, but experts often separate them into three types: Integrated Coastal Management; Integrated Ocean or Seas Management; and Integrated Coastal Zone Management. The difference is significant. Integrated Coastal Management may involve the management

Definition of Integrated Coastal & Ocean Management (ICOM)

ICOM is a multidisciplinary and multi-sectoral process that horizontally integrates units of government, while at the same time vertically integrating levels of government, NGOs and communities in the planning and implementation of a distinct program for the protection, conservation, and sustainable development of coastal and/or ocean environments and resources. The planning and management process is based on the physical, socioeconomic and political inter-connections both within and between dynamic coastal and/or ocean systems.

Integrated coastal management in Europe

Close to 70 percent of the European coastline is highly threatened as a direct or indirect result of human-induced impacts. These include: loss of vital coastal habitats (e.g. wetlands and tidal flats); loss of biodiversity; decline in coastal water quality; and competition for limited space.

In response to increasing population pressures and loss of key coastal environments, the European Union (EU) launched an Integrated Coastal Management Demonstration Program in 1996. Running until 1999, it funded 35 pilot projects in 17 European countries. The objective was to test and evaluate how ICM could be implemented and sustained through seed money, with the aim that member states would see the benefits of implementing management programs.

Unfortunately, when the funding ran out in 1999/2000, most of the pilot efforts withered.

of only the landward side (e.g. Costa Rica) or the seaward side of the coastal zone (e.g. Great Barrier Reef Marine Park), not necessarily both. Integrated Coastal Zone Management requires that the planning and management area include both landward and ocean components, such as, a nation's Exclusive Economic Zone (EEZ), near-shore coastal and estuarine waters, adjoining inter-tidal areas, adjacent near-shore land, and the watersheds of coastal rivers.

Unfortunately, having a management plan and an implementation strategy, does not necessarily translate into a viable and effective program with a strong enabling law, an adequate budget, enforcement powers, and competent personnel. Integrated coastal management suffers from: a lack of government commitment and support; inadequate and overlapping administrative or jurisdictional arrangements; failure to allocate sufficient expertise or funding; failure to win the support of the main stakeholders; and in most developing countries, embedded and pervasive corruption within all levels of government and the private sector.

INTEGRATED COASTAL MANAGEMENT PLANS
2002

National management plans
1
2 – 3
4 – 6

Sub-national, state, or regional plans
1 – 10
44 or more

International Management Plans

In the past three decades there has been a proliferation of international and regional integrated coastal management efforts. Virtually every sea now has at least one regional management program in place.

As of late 2002, 168 countries were participating in one or more regional initiatives aimed at managing shared seas and/or coastal areas. These are comprised of two general groups: 40 regional Integrated Coastal Management programs involving a total of 84 countries; and the United Nations Environment Program's (UNEP) Regional Seas Program which has 13 large-scale regional efforts involving 140 countries.

The reason for the proliferation of regional management strategies and action plans is due, in large measure, to the pioneering efforts of UNEP's Regional Seas Program launched in 1974, as a result of the 1972 UN Conference on the Human Environment in Stockholm. It was followed the same year by the Baltic Sea Joint Comprehensive Environmental Action Program, endorsed by all Baltic States, with a permanent secretariat based in Helsinki.

The early successes of the Regional Seas Program – which launched its flagship Mediterranean Action Plan in 1975 – gave impetus to a flurry of similar efforts by countries sharing common seas and common challenges. The recognition of the inter-related nature of near-shore marine and coastal

UNEP'S REGIONAL SEAS PROGRAM

○ countries party to more than one Action Plan

There are 13 regional seas programs incorporating management plans and some activities in 140 countries. All programs have an agreed action plan; a few also have a regional convention and associated protocols in place.

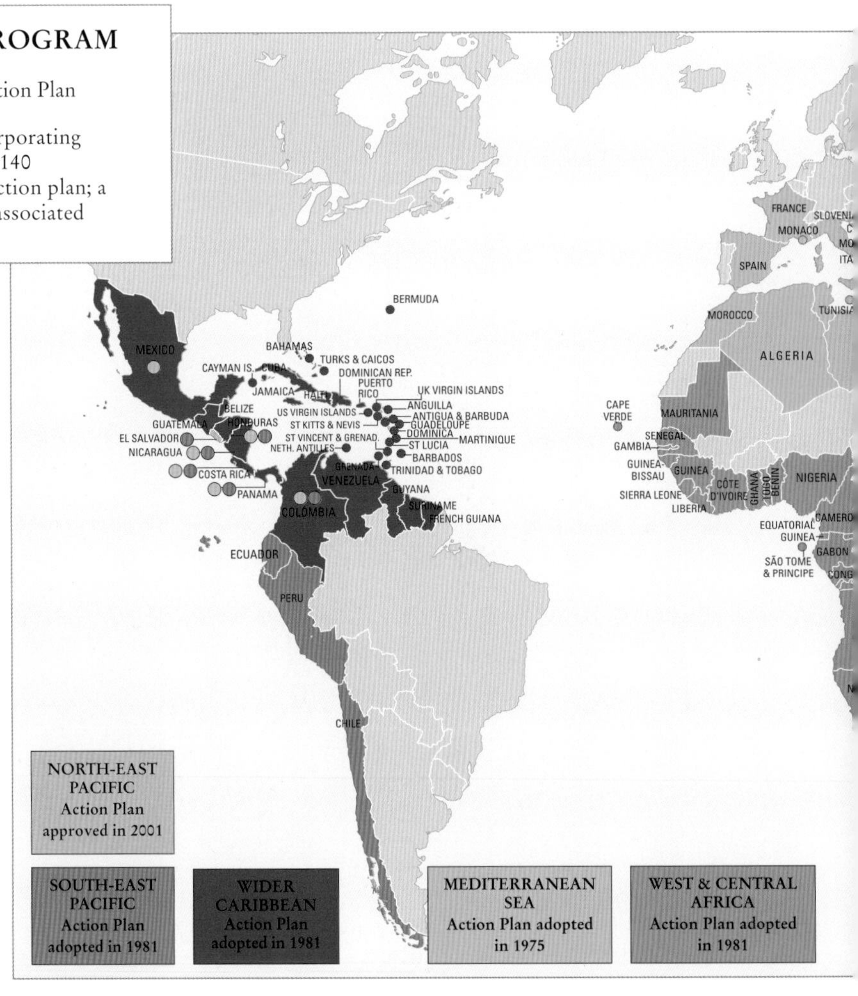

Eight key lessons

The eight key requirements for successful region-wide coastal management programs are:

1) Local government commitment to project implementation;
2) Management awareness of transboundary issues;
3) Good partnerships between implementing agencies;
4) Familiarity with implementation requirements;
5) Indigenous management and expertise input;
6) Good working relationships between government and research/academic institutions;
7) Management boundaries defined in terms of ecosystem features and impacts;
8) Research clearly geared to addressing management issues.

environments, coupled with their vital economic importance as drivers of development and trade, including tourism, prompted many countries to develop international approaches to managing common seas, in addition to national or sub-national efforts.

Achieving real progress in managing the complex interplay between land and sea, including the integrity of coastal ecosystems and the biodiversity they harbor, requires planning and interventions at three broad levels of governance: framework policies governing oceans and coasts at the national level, backed up by local expertise and capacity; locally or regionally oriented site management programs governing specific areas or ecosystems; and the ability to integrate planning and implementation across economic sectors and disciplines.

Unfortunately, only a handful of these regional seas programs are actually functional – the Mediterranean, Caribbean, and Black Sea, and to some extent the Northwest Pacific and South Pacific. The rest have become paper plans, with good intentions and an action plan, but little in the way of implementation.

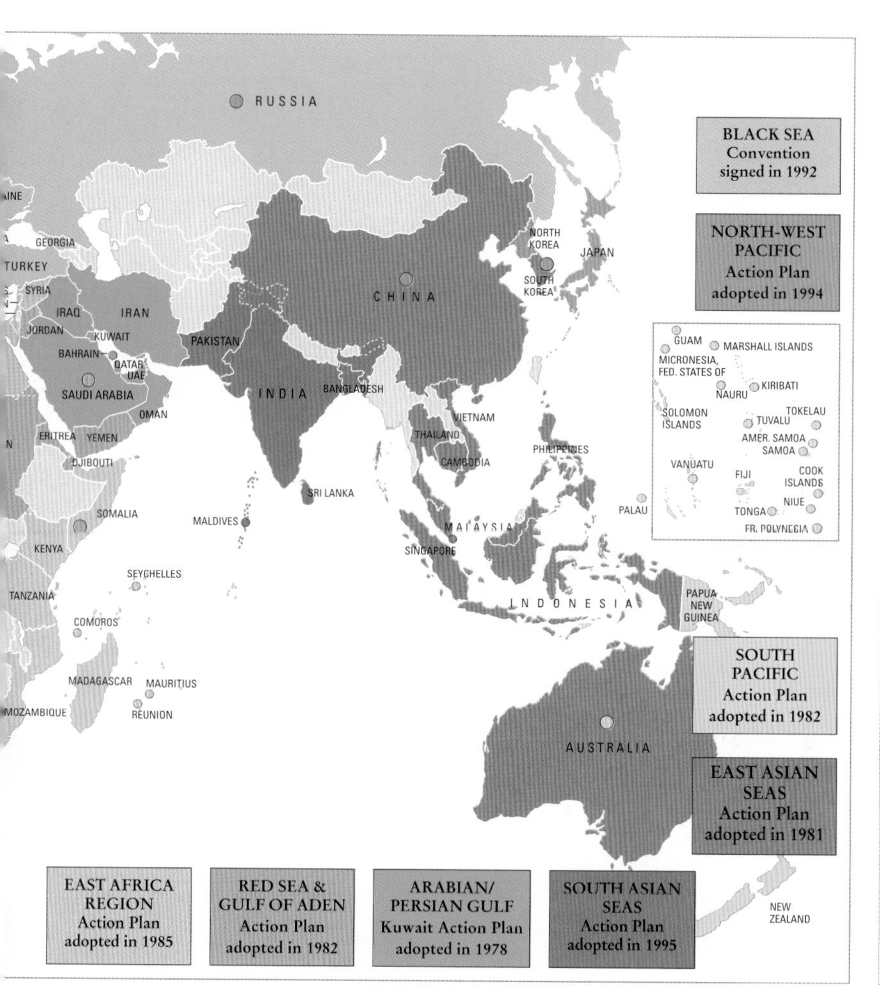

Major international management efforts

Great Lakes Water Quality Management: Canada and US.

Gulf of Maine Program and Action Plan: Canada (New Brunswick and Nova Scotia) and US states (Maine, Massachusetts, New Hampshire).

Puget Sound-Georgia Basin Environmental Initiative: US and Canada.

Tijuana River Watershed and Estuary Project: US and Mexico.

Partnership in Environmental Management for the Seas of East Asia (PEMSEA): Brunei, Cambodia, China, Indonesia, Japan, Malaysia, North Korea, Philippines, Singapore, South Korea, Thailand, and Vietnam (has secretariat but implementation slow).

European Network of Local Estuarine Authorities: France, Portugal, and UK.

Baltic Sea Joint Comprehensive Environmental Action Programme: Denmark, Estonia, Finland, Germany, Latvia, Lithuania, Poland, Russia, and Sweden.

The Convention for the Protection of the Marine Environment of the Northeast Atlantic: Belgium, Denmark, Finland, France, Germany, Iceland, Ireland, Luxembourg, Netherlands, Norway, Portugal, Spain, Sweden, Switzerland, and UK.

North Sea Coastal Zone Management: Belgium, Denmark, Germany, Netherlands, Norway, and UK.

Trilateral Wadden Sea Plan: Denmark, Germany, and Netherlands.

Skagerrak Forum: Denmark, Norway, and Sweden.

Arctic Environmental Protection Strategy: Canada, Denmark (including Greenland & Faroe Islands), Finland, Iceland, Norway, Russia, Sweden, and US.

Convention on the Conservation of Antarctic Marine Living Resources: Argentina, Australia, Belgium, Chile, China, European Union, France, Germany, India, Italy, Japan, South Korea, Namibia, New Zealand, Norway, Poland, Russia, South Africa, Spain, Sweden, Ukraine, United Kingdom, US, and Uruguay.

98–99 ▷▷

International Management Plans: The Mediterranean

In 1975, 16 Mediterranean countries and the European Union adopted the Mediterranean Action Plan (MAP), the first-ever plan adopted under UNEP's Regional Seas Programme. Initially, the main objectives of MAP were to assist Mediterranean countries in: assessing and controlling marine pollution; formulating more integrated national environmental policies; improving the ability of governments to identify more sustainable patterns of development; and optimizing the allocation of resources.

In 1976, the MAP countries adopted the Convention for the Protection of the Mediterranean Sea Against Pollution. Known as the Barcelona Convention, it has spawned seven protocols.

Though originally designed to control marine pollution and maintain critical coastal and marine ecosystems, MAP has evolved over the past 30 years to include integrated coastal zone planning and management. In 1995, the original MAP was replaced by the Action Plan for the Protection of the Marine Environment and the Sustainable Development of the Coastal Areas of the Mediterranean, known as MAP, Phase II.

Currently there are 22 parties to the Barcelona Convention. The expanded action plan focuses on: assessing and controlling marine pollution; ensuring sustainable management of marine and coastal resources; integrating environmental issues into social and economic development; protecting the marine environment from land- and sea-based pollution; protecting the natural and cultural heritage of the region; strengthening coastal management capacity among states; and improving the quality of life for the Mediterranean's population (enshrined in the region's Blue Plan).

Unfortunately, in most Mediterranean states there is an administrative separation between land and sea that impedes the ability of states to manage coastal areas. By 2006 only five Mediterranean states had a framework law for governing coastal areas – Greece, Lebanon, France, Spain, and Algeria. In January 2008, the Integrated Coastal Zone Management Protocol was added to the Barcelona Convention. Once this Protocol enters into force, it will assist Mediterranean states to design and implement integrated coastal zone management programs that use an ecosystem approach to planning and management, apply coastal setbacks of at least 100 meters to prevent unchecked shore construction, introduce sustainable beach management practices, undertake measures to respond to natural disasters, and share information and best practices.

The Mediterranean Sea hosts 7% of known ocean flora and fauna in only 0.3% of the world oceans' volume

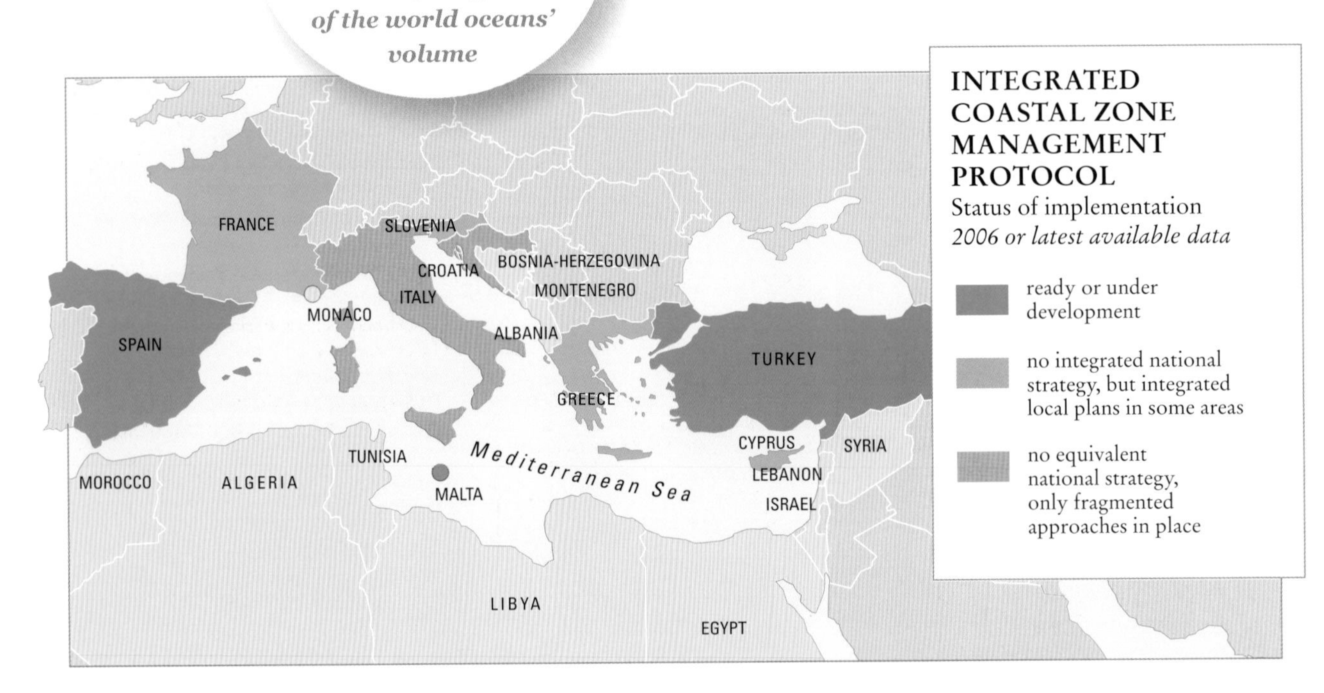

64–65, 96–97

ESTABLISHING SPECIALLY PROTECTED AREAS (SPAS)

The MAP has facilitated the setting up of specially protected areas in order to preserve critical ecosystems and promote species conservation. From just 50 SPAs in 2000, there are now more than 800 covering some some 144,000 square kilometers (55,598 square miles); two-thirds of them in marine areas. There is an additional list of Specially Protected Areas of Mediterranean Interest (SPAMI) that includes areas of importance for preserving the biological diversity of the Mediterranean, or areas that include habitats specific to the region, such as the Mediterranean maquis or *Posidonia* seagrass beds.

Most of these SPAs are located in the western basin of the Mediterranean – 82 percent compared to 18 percent in the eastern basin. There is also a north-south divide, with European states accounting for 712 SPAs, while North Africa and the Middle East account for only 131.

The MAP has spurred the safeguarding of critical habitats and the biodiversity they shelter, putting the region on its way to meeting the international objective of protecting 10 percent of the world's marine ecosystems by 2020.

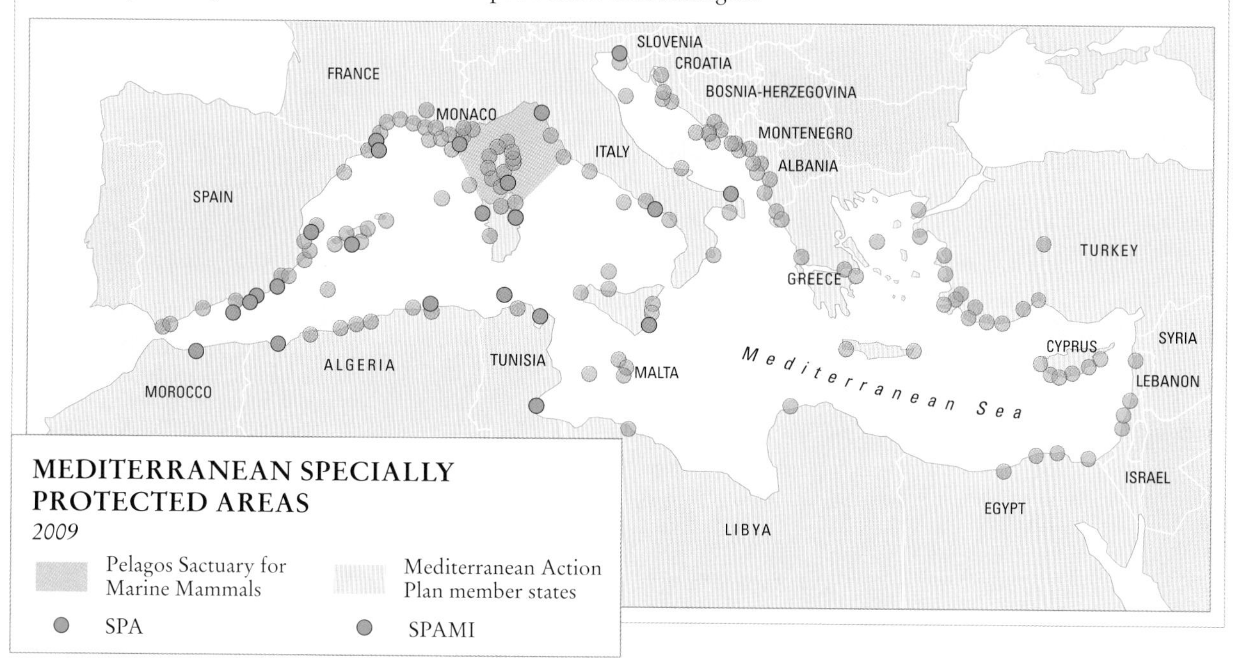

MEDITERRANEAN SPECIALLY PROTECTED AREAS
2009

MAP's seven protocols under the Barcelona Convention

1) Protocol for the Prevention of Pollution in the Mediterranean Sea by Dumping from Ships and Aircraft; entered into force: February 1976; amended June 1995.
2) Protocol Concerning Cooperation in Preventing Pollution from Ships and, in Cases of Emergency, Combating Pollution of the Mediterranean Sea; entered into force: January 2002.
3) Protocol for the Protection of the Mediterranean Sea against Pollution from Land-based Sources and Activities; entered into force: May 2008.
4) Protocol Concerning Specially Protected Areas and Biological Diversity in the Mediterranean; entered into force: December 1999.
5) Protocol for the Protection of the Mediterranean Sea against Pollution Resulting from Exploration and Exploitation of the Continental Shelf and the Seabed and its Subsoil; not yet in force.
6) Protocol on the Prevention of Pollution of the Mediterranean Sea by Transboundary Movements of Hazardous Wastes and their Disposal; entered into force: January 2008
7) Protocol on Integrated Coastal Zone Management in the Mediterranean; not yet in force.

Marine Protected Areas

The number of marine protected areas has increased in recent years, but they still represent less than one percent of the ocean area, woefully inadequate to protect marine ecosystems and biodiversity.

Oceans cover 70% of the planet but only 0.65% are protected while 12% of land is protected

Currently there are around 5,000 marine protected areas throughout the world's seas, nearly 2.6 million square kilometers of the marine environment. Despite the fact that the number of marine reserves and protected areas has increased rapidly over the past 15 years, they still occupy less than one percent of the world's oceans.

Faced with multiple threats – from rampant over-fishing to degradation and outright destruction of vital near-shore and marine ecosystems such as mangrove swamps, seagrass beds and coral reefs – there is an urgent need to conserve and protect marine habitats and the biodiversity they support.

In 2001, 161 of the world's top marine scientists called for urgent action to set up a global network of marine reserves. A team of UK marine scientists calculated that managing a reserve system covering an impressive 30 percent of the world's oceans, would cost a modest US $12–14 billion a year, provide up to a million new jobs and significantly increase the world's oceanic fish catch.

The benefits of conserving marine areas are increasingly evident: when the US state of Florida set up a series of marine protected areas and no-fish zones, scientists noted that within four years densities of yellow-tailed snapper increased more than 15 times, compared to unprotected areas; after Apo Island, off the coast of Negros, Philippines, set up a marine protected area in 1986 comprising just eight percent of the 106 hectare reef that fringed the island, catches of fish rebounded after two years and have since increased ten-fold.

Protecting ocean ecosystems is a sensible and cost effective way to restore fisheries and boost coastal incomes, while maintaining vital ecosystem services such as protecting shorelines, stabilizing sediments, sequestering carbon, and filtering out pollutants.

MARINE PROTECTED AREAS
2010

exclusive economic zones

marine protected areas

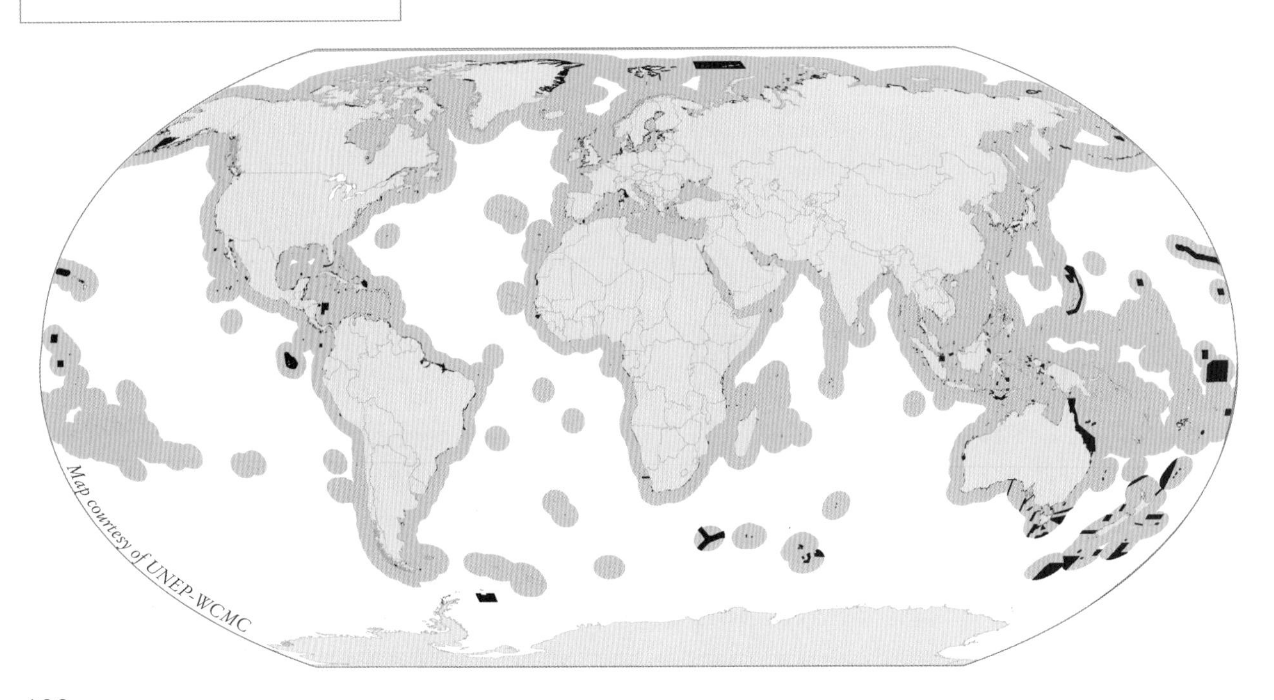

Turtle conservation

During the past two decades leatherback turtles have suffered from egg collectors robbing their nests and fishing operations snagging turtles as a by-catch while trawling for fish, resulting in a 90 percent reduction in numbers. Satellite tracking of leatherbacks in the eastern Pacific from 2004–07 recorded over 12,000 days of travel and the specific routes taken across the Pacific from Costa Rica, across the equator and into the South Pacific Gyre. As a result, fisheries have been closed during migrations, and using the tracking data, Costa Rica has been able to protect turtle nesting sites in the marine Caletas Wildlife Refuge and improve zoning in their EEZ, measures which are having a positive impact on conservation of the species.

AUSTRALIA'S GREAT BARRIER REEF MARINE PARK: A MODEL FOR MANAGEMENT

The Great Barrier Reef Marine Park stretches more than 2,000 kilometers (1,242 miles) along the Queensland coast, encompassing 350,000 square kilometers (135,135 square miles), an area larger than the UK. Divided into four main management zones, each zone limits activities according to three types of use – general use zones, national park zones and preservation or scientific research zones. The Park's 70 regions contain 2,900 separate reef formations, 760 fringing reefs, 300 reef islands or cays and 600 near shore continental islands.

In 2004 all commercial and recreational fishing was banned in 32 percent of the Park's area. The biomass of many fish species, including coral trout and many species of groupers, doubled within two years. In addition, there are far fewer infestations of the coral eating crown-of-thorns starfish. The Park is host to dazzling biodiversity and generates over US $3.4 billion a year in tourist revenue.

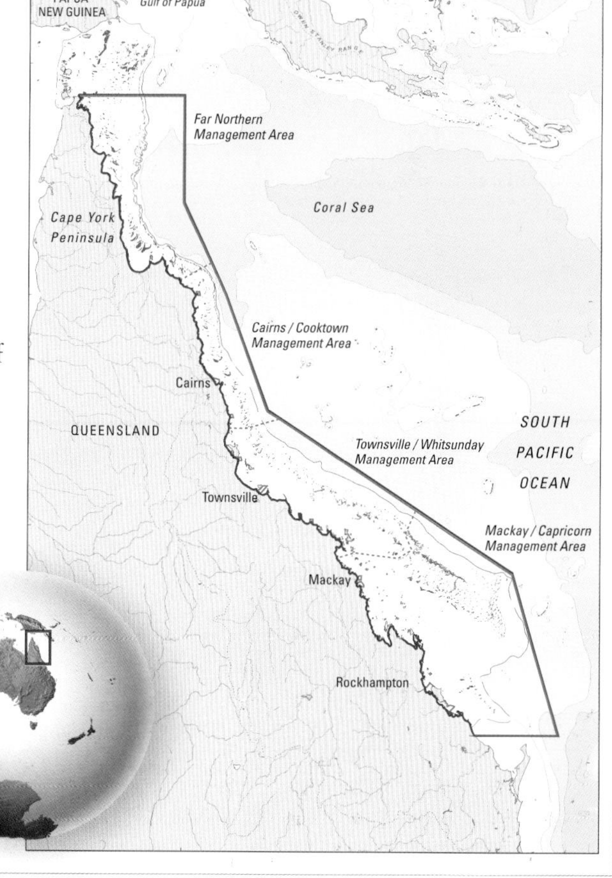

GREAT BARRIER REEF MARINE PARK
2009

- World Heritage Area and Region boundary
- Great Barrier Reef Marine Park, total area 344,400 km²
- Great Barrier Reef Marine Park management areas

Large Marine Ecosystems (LMEs)

There are 64 Large Marine Ecosystems encompassing all coastal areas of the oceans and major current systems. Together they account for the majority of the world's fish catches, as well as receiving most of the pollution entering the seas.

Large Marine Ecosystems (LMEs), in theory, provide a flexible approach to ecosystem-based management of large coastal and ocean areas by identifying and addressing the main drivers of ecosystem change. The US National Oceanographic and Atmospheric Administration (NOAA) has supported the development of LME management and assessment strategies since 1984. Since then, NOAA has partnered with the Intergovernmental Oceanographic Commission (IOC) of UNESCO, IUCN, and various UN agencies, including UNDP, UNEP, and the World Bank's Global Environment Facility (GEF).

As of 2009, 121 countries were in the process of meeting ecosystem-related targets in efforts to address a number of crucial areas of concern, including rampant over-fishing, non-sustainable exploitation of food webs, destruction of key habitats, and accelerated nitrogen (nutrient) pollution, among others. Additionally, 111 countries were engaged in transboundary diagnostic analysis in order to identify the primary causes of loss of marine biomass, coastal pollution, damaged habitats and depleted fish stocks.

The Humboldt Current LME

The Humboldt Current off the west coast of South America is unique: it contains the world's largest upwelling system and therefore is the most productive marine ecosystem on the planet. It provides 15 to 20 percent of the world's total annual marine catch. Fishing sustains tens of thousands of fishers and their families, with anchovies, sardines, and mackerel comprising the backbone of commercial fisheries.

The high productivity of the current is the result of upwelling processes governed, in turn, by strong southerly trade winds. These are, however, subject to seasonal variations and are interrupted by the El Niño Southern Oscillation which results in the intrusion of warmer surface waters and a marked reduction in fish catches.

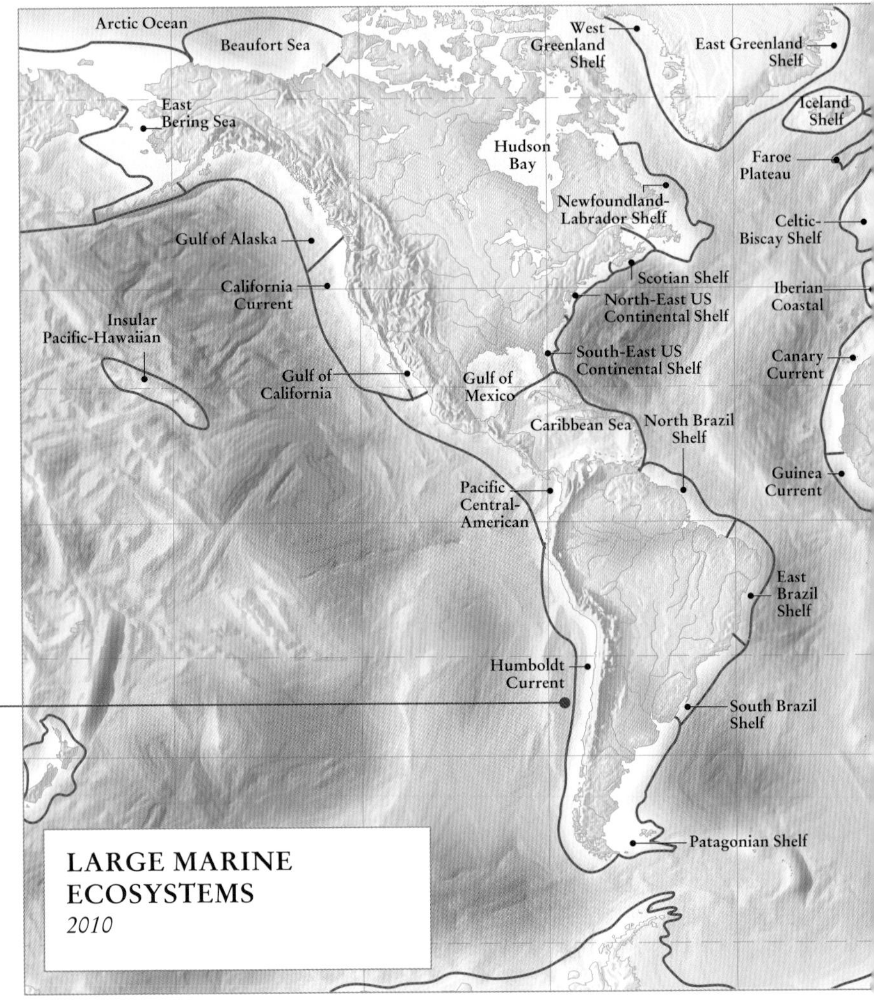

NOAA and IUCN are currently involved in a large-scale joint program designed to assist developing countries in planning and implementing an ecosystem-based strategy to assess and manage coastal and ocean resources. In a complementary initiative, the IOC of UNESCO and its partners – World Meteorological Organization, UNEP and the International Council for Science – are working together to develop relevant ecosystem monitoring and forecasting methods that can be applied to both marine and coastal areas.

The LME approach to assessment and management is intended to allow groups of countries to improve their capacity for the conservation and sustainable development of marine and coastal resources. Unfortunately, as with most other integrated coastal management efforts, the work remains on paper. Only five of 64 – the Mediterranean, Black Sea, Baltic, Caribbean, and North Seas – have implementation mechanisms in place. Many of the others have been set up to manage fisheries only – not the actual ecosystems they are dependent upon.

LMEs produce 95 % of the world's marine fish catch

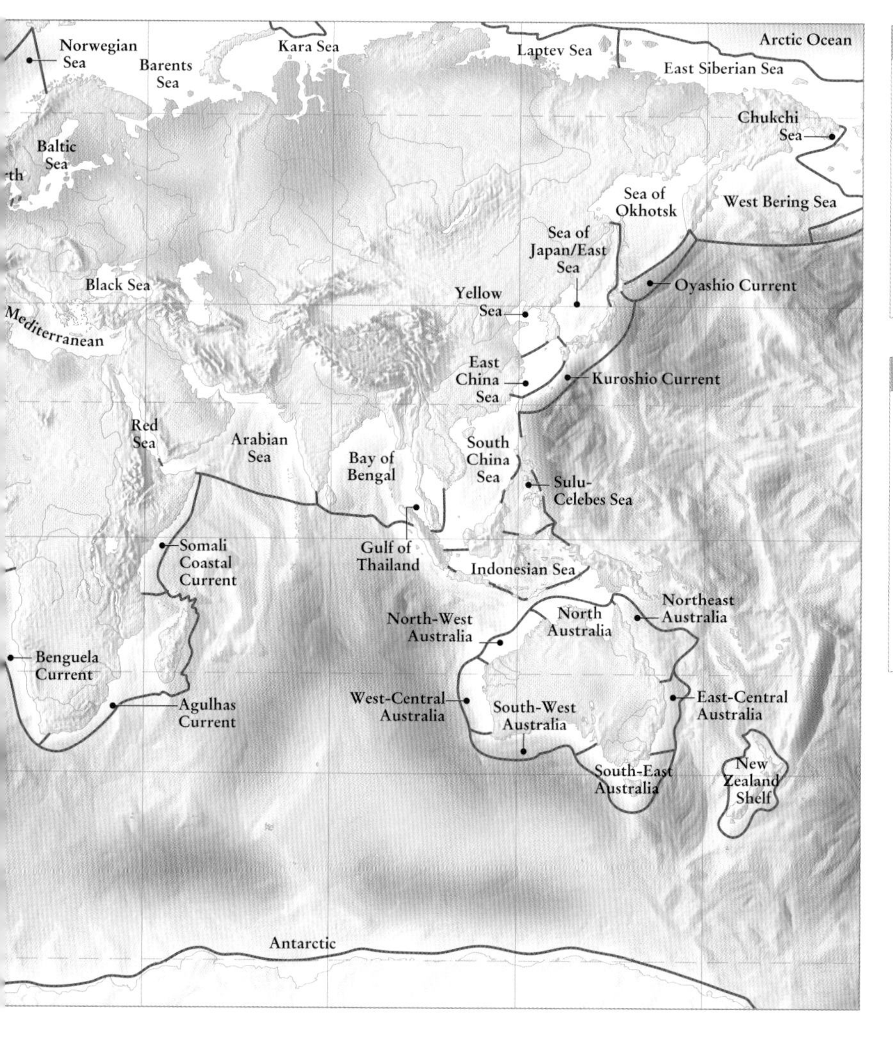

Definition of LMEs

LMEs are defined by four ecological criteria:

1) bottom depth contours (bathymetry);
2) currents and water mass (hydrography);
3) marine productivity;
4) food webs (trophic relationships).

Composition of LMEs

LMEs are 200,000 square kilometers (77,220 square miles) or more in size. They encompass ocean or coastal areas, including river basins and estuaries, seaward to the continental shelf (e.g. the Yellow Sea) or out to the extent of a specific current system (e.g. the Guinea Current off the West Coast of Africa).

Marine Ecosystems and Species in Peril

Our collective failure to act to protect the seas leaves many of the world's marine ecosystems and species critically endangered. For some, such as corals, any action now may well be too late. Hundreds of countries have integrated management plans that remain on paper, very few of them are currently under active implementation. Yet continued inaction endangers not only the many other species that live in or depend on the marine environment but also the people who are dependant on the coasts and oceans. The future for these millions – marine and human – is indeed perilous.

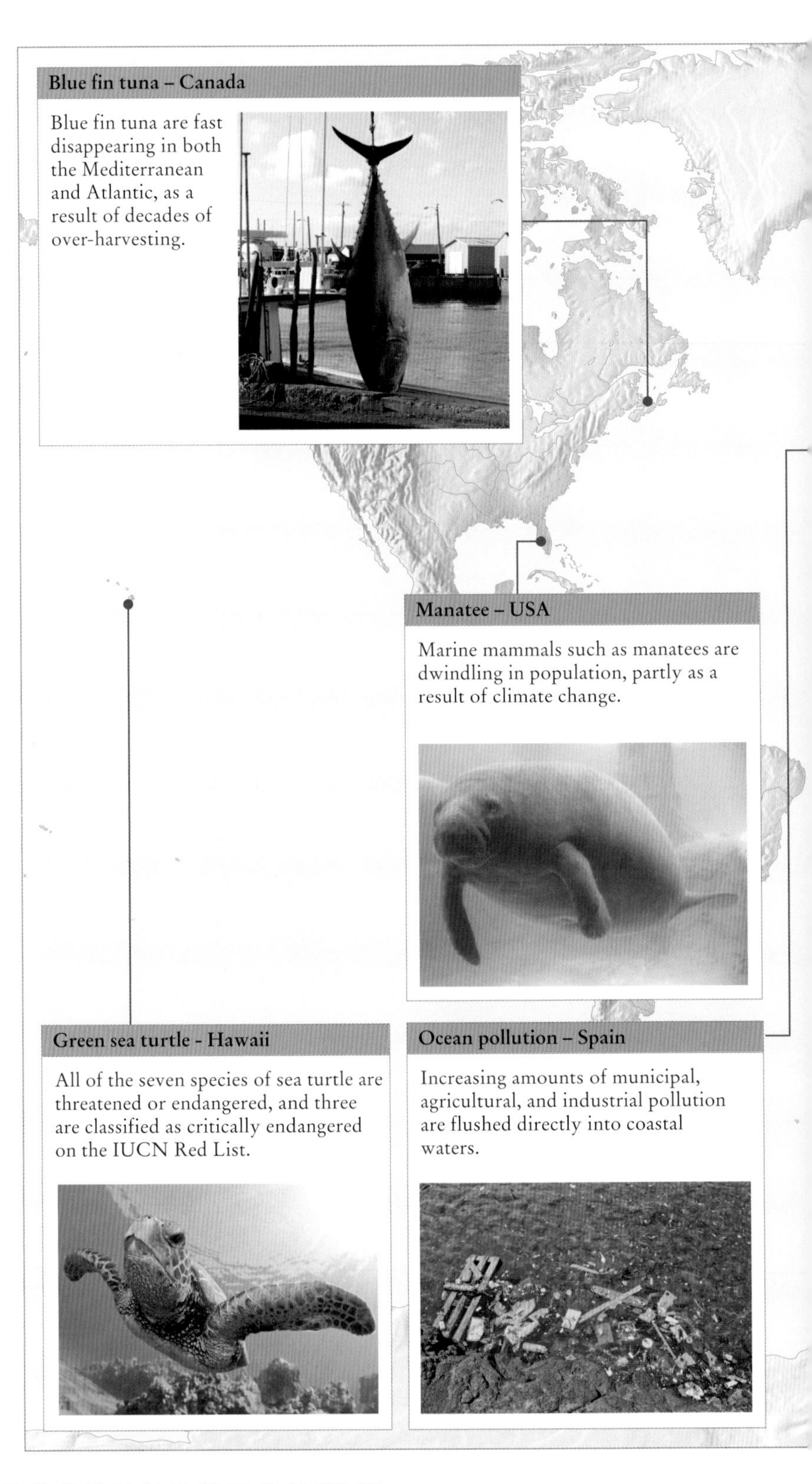

34–35, 36–37, 44–45, 46–47, 48–49, 50–51, 74–75, 80–81, 82–83, 100–101

Polar bear – Svalbard

Sea ice is being reduced by global warming, putting polar bears and other polar mammals at risk. The Center for Biological Diversity warns that polar bear numbers could be reduced two-thirds by 2050.

Soft coral – Indonesia

Corals are doubly at risk – endangered by the rising surface water temperatures caused by global warming which can trigger coral bleaching, ending in the death of coral polyps, the animals that build reefs, and by ocean acidification.

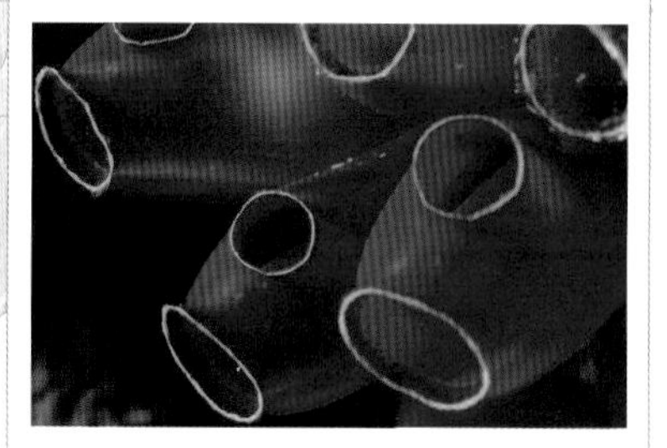

Seagrasses – The Mediterranean

Seagrass communities provide feeding, breeding, and nursery habitats for thousands of marine species, yet are more threatened than tropical rainforests.

Mangrove – Indonesia

The IUCN considers 16% of mangrove species worldwide to be in danger of extinction, and 40% to be at risk.

Anchovies – The Mediterranean

80% of commercial fisheries worldwide are being exploited at non-sustainable rates.

Sea urchin – South Africa

Ocean acidification affects the ability of phytoplankton, molluscs, and echinoderms, such as sea urchins, to form their shells or skeletons.

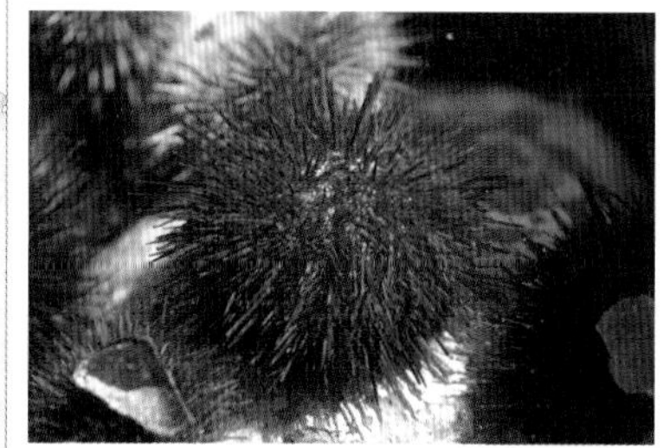

Flooding – Tuvalu

Tuvalu is at risk of inundation by rising sea levels and increasingly frequent storm surges.

	1 Total population 1,000s		2 Urban population as % of total		3 Population within 100 km of coast as % of total population	4 Length of coastline km	5 Area of claimed EEZ km²	6 Mangrove forest area km²
	2010	*projected 2025*	*2010*	*projected 2025*	*2000*	*2010*	*2000*	*1997*
Afghanistan	29,117	44,970	23%	29%	0%	0	–	–
Albania	3,169	3,395	52%	64%	97%	362	–	–
Algeria	35,423	42,882	66%	74%	69%	998	–	–
American Samoa	69	86	93%	95%	100%	116	–	–
Angola	18,993	27,441	59%	69%	29%	1,600	–	1,250.00
Anguilla	15	18	100%	100%	100%	61	–	–
Antigua and Barbuda	89	101	30%	35%	100%	153	102,867	13.2
Argentina	40,666	45,883	92%	94%	45%	4,989	925,362	–
Armenia	3,090	3,181	64%	67%	0%	0	–	–
Aruba	107	112	47%	50%	100%	69	–	4.2
Australia	21,512	24,703	89%	91%	90%	25,760	6,664,107	11,500.00
Austria	8,387	8,600	68%	72%	2%	0	–	–
Azerbaijan	8,934	10,128	52%	56%	56%	0	–	–
Bahamas	346	402	84%	87%	100%	3,542	369,149	2,332.00
Bahrain	807	1,021	89%	90%	100%	161	–	1
Bangladesh	164,425	195,012	28%	37%	55%	580	39,868	5,767.00
Barbados	257	262	44%	54%	100%	97	183,436	< 0.1
Belarus	9,588	8,851	75%	81%	0%	0	–	–
Belgium	10,698	11,191	97%	98%	83%	67	–	–
Belize	313	404	52%	60%	100%	386	12,839	719
Benin	9,212	13,767	42%	50%	62%	121	–	17
Bermuda	65	66	100%	100%	100%	103	446,869	0.1
Bhutan	708	865	35%	46%	0%	0	–	–
Bolivia	10,031	12,368	67%	73%	0%	0	–	–
Bosnia & Herzegovina	3,760	3,608	49%	58%	47%	20	–	–
Botswana	1,978	2,337	61%	70%	0%	0	–	–
Brazil	195,423	213,802	87%	90%	49%	7,491	3,442,548	13,400.00
Brunei	407	513	76%	81%	100%	161	5,614	171
Bulgaria	7,497	6,752	71%	76%	29%	354	25,699	–
Burkina Faso	16,287	24,837	26%	39%	0%	0	–	–
Burma	50,496	57,585	34%	44%	49%	1,930	358,495	3,786.00
Burundi	8,519	11,161	11%	17%	0%	0	–	–
Cambodia	15,053	18,973	20%	26%	24%	443	–	851
Cameroon	19,958	26,478	58%	68%	22%	402	10,900	2,494.00
Canada	33,890	38,659	81%	83%	24%	202,080	3,006,236	–
Cape Verde	513	616	61%	70%	100%	965	742,438	–
Cayman Islands	57	63	100%	100%	100%	160	–	–
Central African Republic	4,506	5,747	39%	45%	0%	0	–	–
Chad	11,506	16,906	28%	38%	0%	0	–	–
Channel Islands	150	152	31%	36%	100%	–	–	–
Chile	17,135	19,266	89%	92%	82%	6,435	3,415,864	–
China	1,354,146	1,453,140	47%	59%	24%	14,500	–	366
China, Hong Kong SAR	7,069	7,969	100%	100%	100%	733	–	2.8
China, Macao SAR	548	603	100%	100%	100%	41	–	–
Colombia	46,300	54,920	75%	80%	30%	3,208	706,134	3,659.00

7 **Marine Protected areas** number *2008*	8 **Fisheries capture** Tons of fish, shellfish, and molluscs captured *2008*	9 **Aquaculture** Tons of fish, shellfish and molluscs produced *2008*	10 **Fish species threatened, vulnerable or rare** number *2008*	11 **Improved sanitation** % of population with access *2008*	12 **Organic water pollutant (BOD) emissions** kg per day *2005*	
0	1,000	-	3	37%	-	Afghanistan
7	5,510	1,858	33	98%	3,349	Albania
6	138,833	2,781	23	95%	–	Algeria
7	4,451	-	8	–	–	American Samoa
4	317,262	-	22	57%	–	Angola
–	701	-	–	99%	–	Anguilla
12	3,521	–	14	–	–	Antigua and Barbuda
36	995,083	2,700	31	90%	-	Argentina
0	3,700	2,001	4	90%	–	Armenia
1	151	-	15	–	-	Aruba
384	178,576	57,152	84	100%	–	Australia
0	350	2,087	9	100%	84,769	Austria
0	1,517	89	9	45%	18,107	Azerbaijan
32	9,117	0	20	100%	-	Bahamas
2	14,177	2	6	–	–	Bahrain
7	1,557,754	1,005,542	12	53%	-	Bangladesh
3	3,551	-	15	100%	–	Barbados
0	900	4,150	1	93%	–	Belarus
2	22,609	126	9	100%	97,883	Belgium
19	4,621	9,549	22	90%	–	Belize
0	37,495	180	15	12%	–	Benin
101	400	-	12	–	–	Bermuda
0	180	0	0	65%	–	Bhutan
0	6,797	631	0	25%	-	Bolivia
0	2,005	7,620	27	95%	–	Bosnia & Herzegovina
0	86	-	2	60%	3,440	Botswana
58	775,000	290,186	64	80%	–	Brazil
7	2,358	473	8	–	–	Brunei
1	8,861	5,157	17	100%	100,634	Bulgaria
0	10,600	405	0	11%	–	Burkina Faso
6	2,493,750	674,776	17	81%	–	Burma
0	17,766	200	18	46%	–	Burundi
2	431,000	40,000	18	29%	-	Cambodia
2	138,000	340	43	47%	–	Cameroon
563	937,370	144,099	26	100%	-	Canada
1	21,910	-	18	54%	–	Cape Verde
41	125	-	14	96%	–	Cayman Islands
0	15,000	0	0	34%	–	Central African Republic
0	40,000	-	0	9%	–	Chad
–	3,228	972	–	–	–	Channel Islands
9	3,554,814	843,142	18	96%	95,243	Chile
36	14,791,163	32,735,944	70	55%	–	China
22	158,126	66,400	13	–	–	China, Hong Kong SAR
–	1,500	4,754	6	–	–	China, Macao SAR
15	135,000	-	31	74%	86,992	Colombia

	1 Total population 1,000s		2 Urban population as % of total		3 Population within 100 km of coast as % of total population	4 Length of coastline km	5 Area of claimed EEZ km²	6 Mangrove forest area km²
	2010	*projected 2025*	*2010*	*projected 2025*	*2000*	*2010*	*2000*	*1997*
Comoros	691	907	28%	33%	100%	340	161,993	26.2
Congo	3,759	5,094	62%	69%	25%	169	–	120
Congo, dem rep	67,827	98,123	35%	46%	3%	37	–	226
Cook Islands	20	21	75%	83%	–	120	–	–
Costa Rica	4,640	5,521	64%	72%	100%	1,290	542,060	370
Côte d'Ivoire	21,571	29,738	51%	61%	40%	515	157,379	644
Croatia	4,410	4,254	58%	64%	38%	5,835	–	–
Cuba	11,204	11,148	75%	76%	100%	3,735	222,204	7,848.00
Cyprus	880	1,014	70%	74%	100%	648	–	–
Czech Republic	10,411	10,573	74%	76%	0%	0	–	–
Denmark	5,481	5,590	87%	89%	100%	7,314	80,419	–
Djibouti	879	1,111	76%	79%	100%	314	2,488	10
Dominica	67	68	67%	71%	100%	148	24,917	1.6
Dominican Republic	10,225	11,973	69%	77%	100%	1,288	246,454	325
East Timor	1,171	1,869	28%	36%	–	706	–	–
Ecuador	13,775	16,074	67%	75%	61%	2,237	–	2,469.00
Egypt	84,474	104,970	43%	48%	53%	2,450	185,304	861
El Salvador	6,194	6,895	64%	72%	99%	307	–	268
Equatorial Guinea	693	971	40%	46%	72%	296	291,445	277
Eritrea	5,224	7,404	22%	31%	74%	2,234	–	581
Estonia	1,339	1,321	69%	72%	86%	3,794	11,558	–
Ethiopia	84,976	119,822	17%	21%	1%	0	–	–
Falkland Islands (Malvinas)	3	3	74%	80%	–	1,288	–	–
Faroe Islands	50	55	40%	44%	–	1,117	–	–
Fiji	854	905	52%	59%	100%	1,129	1,055,048	385
Finland	5,346	5,533	85%	88%	73%	1,250	–	–
France	62,637	65,769	85%	91%	40%	3,427	706,443	–
French Guiana	231	323	76%	80%	93%	378	131,640	55
French Polynesia	272	318	51%	54%	80%	2,525	4,553,115	–
Gabon	1,501	1,915	86%	90%	63%	885	180,676	2,500.00
Gambia	1,751	2,478	58%	68%	91%	80	20,530	497
Georgia	4,219	3,888	53%	56%	39%	310	18,876	–
Germany	82,057	79,258	74%	77%	15%	2,389	37,438	–
Ghana	24,333	32,233	51%	62%	43%	539	216,867	100
Gibraltar	31	32	100%	100%	–	12	–	–
Greece	11,183	11,274	61%	67%	99%	13,676	–	–
Greenland	57	56	84%	88%	100%	44,087	–	–
Grenada	104	109	39%	48%	100%	121	20,285	2.4
Guadeloupe	467	489	98%	99%	100%	306	115,679	39.8
Guam	180	211	93%	94%	–	126	–	–
Guatemala	14,377	19,927	49%	58%	61%	400	104,530	161
Guinea	10,324	15,158	35%	45%	41%	320	96,973	2,963.00
Guinea-Bissau	1,647	2,296	30%	35%	95%	350	86,670	2,484.00
Guyana	761	732	29%	34%	77%	459	122,017	800
Haiti	10,188	12,476	52%	69%	100%	1,771	86,398	134

7 Marine Protected areas number 2008	8 Fisheries capture Tons of fish, shellfish, and molluscs captured 2008	9 Aquaculture Tons of fish, shellfish and molluscs produced 2008	10 Fish species threatened, vulnerable or rare number 2008	11 Improved sanitation % of population with access 2008	12 Organic water pollutant (BOD) emissions kg per day 2005	
1	16,000	-	7	36%	–	Comoros
0	54,104	65	15	30%	–	Congo
1	236,000	2,970	25	23%	–	Congo, dem rep
–	3,000	0	–	100%	–	Cook Islands
35	21,750	27,035	19	95%	–	Costa Rica
3	58,000	1,290	19	23%	–	Côte d'Ivoire
19	49,024	12,017	46	99%	41,209	Croatia
42	27,856	33,039	28	91%	–	Cuba
5	2,011	3,403	12	100%	8,000	Cyprus
0	4,164	20,395	5	98%	148,863	Czech Republic
52	690,202	35,337	13	100%	60,522	Denmark
1	450	-	14	56%	–	Djibouti
2	694	0	15	–	–	Dominica
15	15,424	980	15	83%	-	Dominican Republic
0	3,125	51	5	50%	–	East Timor
3	434,239	172,120	15	92%	44,748	Ecuador
8	373,815	693,815	24	94%	-	Egypt
1	48,000	3,766	7	87%	–	El Salvador
4	5,400	0	13	–	–	Equatorial Guinea
0	1,665	-	14	14%	2,871	Eritrea
3	101,037	813	4	95%	16,458	Estonia
0	16,770	0	2	12%	24,137	Ethiopia
–	81,708	0	–	–	–	Falkland Islands (Malvinas)
–	495,348	45,929	–	–	–	Faroe Islands
19	48,453	228	11	–	-	Fiji
15	158,399	13,439	5	100%	61,566	Finland
64	457,127	237,833	31	100%	578,173	France
–	3,957	0	–	–	–	French Guiana
8	11,909	44	13	98%	–	French Polynesia
5	30,000	124	21	33%	–	Gabon
6	42,645	0	16	67%	–	Gambia
2	26,512	180	12	95%	–	Georgia
21	229,499	43,977	20	100%	954,219	Germany
0	349,831	5,594	17	13%	-	Ghana
–	–	–	–	–	–	Gibraltar
12	88,971	114,888	62	98%	-	Greece
2	233,754	-	6	–	–	Greenland
1	2,384	0	15	97%	–	Grenada
–	10,100	33	–	–	–	Guadeloupe
12	302	162	9	99%	–	Guam
7	22,826	18,727	16	81%	–	Guatemala
0	74,000	0	19	19%	–	Guinea
4	6,750	-	18	21%	–	Guinea-Bissau
0	42,168	292	22	81%	–	Guyana
0	10,000	-	15	17%	-	Haiti

	1 Total population 1,000s		2 Urban population as % of total		3 Population within 100 km of coast as % of total population	4 Length of coastline km	5 Area of claimed EEZ km²	6 Mangrove forest area km²
	2010	*projected 2025*	*2010*	*projected 2025*	*2000*	*2010*	*2000*	*1997*
Honduras	7,616	9,844	52%	61%	66%	820	201,203	1,458.00
Hungary	9,973	9,647	68%	74%	0%	0	–	–
Iceland	329	384	93%	95%	100%	4,970	678,708	–
India	1,214,464	1,431,272	30%	37%	26%	7,000	2,103,415	6,700.00
Indonesia	232,517	263,287	44%	51%	96%	54,716	2,914,978	42,550.00
Iran	75,078	87,134	71%	78%	24%	2,440	129,700	207
Iraq	31,467	44,692	66%	68%	6%	58	–	–
Ireland	4,589	5,370	62%	68%	100%	1,448	–	–
Isle of Man	80	80	51%	52%	–	160	–	–
Israel	7,285	8,769	92%	93%	97%	273	–	–
Italy	60,098	60,018	68%	73%	79%	7,600	–	–
Jamaica	2,730	2,866	52%	55%	100%	1,022	234,780	106
Japan	126,995	120,793	67%	71%	96%	29,751	3,648,393	4
Jordan	6,472	8,088	79%	81%	29%	26	–	–
Kazakhstan	15,753	17,025	59%	64%	4%	0	–	–
Kenya	40,863	57,573	22%	30%	8%	536	104,056	530
Kiribati	100	123	44%	49%	99%	1,143	3,387,648	–
Korea, North	23,991	25,128	60%	64%	93%	2,495	72,755	–
Korea, south	48,501	49,484	83%	87%	100%	2,413	202,585	–
Kuwait	3,051	3,988	98%	99%	100%	499	–	–
Kyrgyzstan	5,550	6,378	35%	38%	0%	0	–	–
Laos	6,436	8,273	33%	49%	6%	0	–	–
Latvia	2,240	2,101	68%	69%	75%	498	15,633	–
Lebanon	4,255	4,736	87%	89%	100%	225	–	–
Lesotho	2,084	2,306	27%	38%	0%	0	–	–
Liberia	4,102	5,858	48%	55%	58%	579	–	190
Libya	6,546	8,144	78%	82%	79%	1,770	222,400	–
Lithuania	3,255	2,985	67%	70%	23%	90	3,645	–
Luxembourg	492	582	85%	88%	0%	0	–	–
Macedonia	2,043	2,037	59%	64%	14%	0	–	–
Madagascar	20,146	28,595	30%	38%	55%	4,828	1,079,672	3,403.00
Malawi	15,692	23,194	20%	29%	0%	0	–	–
Malaysia	27,914	33,770	72%	81%	98%	4,675	198,173	6,424.00
Maldives	314	384	40%	56%	81%	644	870,623	–
Mali	13,323	18,603	36%	48%	0%	0	–	–
Malta	410	426	95%	96%	100%	197	–	–
Marshall Islands	63	79	72%	77%	53%	370	1,877,282	–
Martinique	406	418	89%	89%	100%	350	10,615	15.9
Mauritania	3,366	4,443	41%	48%	40%	754	141,334	1
Mauritius	1,297	1,400	42%	45%	100%	177	1,274,638	–
Mayotte	199	277	50%	53%	–	185	–	–
Mexico	110,645	123,366	78%	82%	29%	9,330	2,997,679	5,315.00
Micronesia (Fed. States of)	111	122	23%	27%	98%	6,112	2,906,416	86
Moldova	3,576	3,291	47%	58%	9%	0	–	–
Mongolia	2,701	3,134	62%	69%	0%	0	–	–

7 Marine Protected areas number 2008	8 Fisheries capture Tons of fish, shellfish, and molluscs captured 2008	9 Aquaculture Tons of fish, shellfish and molluscs produced 2008	10 Fish species threatened, vulnerable or rare number 2008	11 Improved sanitation % of population with access 2008	12 Organic water pollutant (BOD) emissions kg per day 2005	
22	12,904	47,080	19	71%	–	Honduras
0	7,394	15,687	9	100%	115,075	Hungary
9	1,284,034	5,098	12	100%	–	Iceland
117	4,104,877	3,478,690	40	31%	–	India
139	4,957,098	1,690,121	111	52%	764,028	Indonesia
12	407,842	154,979	21	–	160,776	Iran
0	34,472	19,246	6	73%	-	Iraq
12	205,342	57,210	16	99%	34,146	Ireland
–	2,770	-	3	–	–	Isle of Man
13	3,435	21,612	31	100%	-	Israel
58	235,785	181,469	33	–	475,760	Italy
12	13,175	5,948	15	83%	–	Jamaica
135	4,248,697	732,374	40	100%	1,122,694	Japan
1	500	540	14	98%	27,112	Jordan
0	55,581	321	13	97%	-	Kazakhstan
11	133,286	4,452	71	31%	–	Kenya
14	34,300	5	7	–	–	Kiribati
0	205,000	63,700	8	–	–	Korea, North
6	1,943,870	473,794	14	100%	316,969	Korea, south
5	4,373	360	10	100%	–	Kuwait
0	8	92	3	93%	11,513	Kyrgyzstan
0	26,925	78,000	6	53%	-	Laos
1	157,934	584	6	78%	29,931	Latvia
1	3,811	803	15	–	-	Lebanon
0	50	91	1	29%	13,208	Lesotho
1	7,890	0	19	17%	–	Liberia
4	47,645	240	14	97%	–	Libya
3	182,763	3,008	6	–	42,872	Lithuania
0	–	–	1	100%	3,830	Luxembourg
0	122	1,331	14	89%	–	Macedonia
8	120,464	9,581	75	11%	88,887	Madagascar
0	70,019	1,700	101	56%	-	Malawi
147	1,395,942	243,081	49	96%	208,441	Malaysia
0	133,086	-	12	98%	–	Maldives
0	100,000	821	1	36%	–	Mali
5	1,279	1,692	13	100%	4,232	Malta
2	35,436	-	10	73%	–	Marshall Islands
–	6,200	0	–	–	–	Martinique
3	195,328	-	23	26%	–	Mauritania
18	6,152	246	11	91%	351	Mauritius
2	12,677	88	3	–	–	Mayotte
38	1,588,857	151,065	114	85%	-	Mexico
15	21,699	0	13	–	–	Micronesia (Fed. States of)
0	1,407	4,700	9	79%	22,390	Moldova
0	88	-	1	50%	–	Mongolia

	1 Total population 1,000s		2 Urban population as % of total		3 Population within 100 km of coast as % of total population	4 Length of coastline km	5 Area of claimed EEZ km²	6 Mangrove forest area km²
	2010	*projected 2025*	*2010*	*projected 2025*	*2000*	*2010*	*2000*	*1997*
Montenegro	626	633	61%	64%	8%	294	–	–
Montserrat	6	7	14%	19%	–	40	–	–
Morocco	32,381	37,865	58%	67%	65%	1,835	328,421	–
Mozambique	23,406	31,190	38%	50%	59%	2,470	493,672	925
Namibia	2,212	2,810	38%	48%	5%	1,572	536,805	–
Nauru	10	11	100%	100%	100%	30	–	–
Nepal	29,853	38,031	19%	28%	0%	0	–	–
Netherlands	16,653	17,348	83%	88%	93%	451	–	–
Netherlands Antilles	201	210	93%	95%	100%	364	–	11.4
New Caledonia	254	304	57%	60%	100%	2,254	1,347,964	456
New Zealand	4,303	4,831	86%	87%	100%	15,134	3,887,441	287
Nicaragua	5,822	7,058	57%	63%	72%	910	–	1,718.00
Niger	15,891	27,388	17%	21%	0%	0	–	–
Nigeria	158,259	210,057	50%	60%	26%	853	164,054	10,515.00
Niue	1	1	38%	46%	–	64	–	–
Northern Mariana Islands	88	111	91%	93%	–	1,482	–	–
Norway	4,855	5,365	79%	84%	95%	25,148	1,095,065	–
Oman	2,905	3,782	73%	77%	89%	2,092	487,356	20
Pakistan	184,753	246,286	36%	43%	9%	1,046	201,520	1,683.00
Palau	21	23	83%	91%	–	1,519	–	–
Palestinian Territories	4,409	6,553	74%	78%	–	0	–	–
Panama	3,508	4,267	75%	82%	100%	2,490	274,600	1,814.00
Papua New Guinea	6,888	9,265	13%	16%	61%	5,152	1,613,759	5,399.00
Paraguay	6,460	8,026	61%	70%	0%	0	–	–
Peru	29,496	34,528	77%	82%	57%	2,414	–	51
Philippines	93,617	117,270	49%	55%	100%	36,289	293,808	1,607.00
Poland	38,038	36,964	61%	63%	14%	440	19,426	–
Portugal	10,732	10,706	61%	69%	93%	1,793	1,656,402	–
Puerto Rico	3,998	4,176	99%	100%	100%	501	187,935	92
Qatar	1,508	1,848	96%	97%	100%	563	–	< 5.0
Réunion	837	973	94%	96%	100%	207	309,956	–
Romania	21,190	19,961	57%	66%	6%	225	18,046	–
Russian Federation	140,367	132,345	73%	76%	15%	37,653	6,255,799	–
Rwanda	10,277	14,676	19%	25%	0%	0	–	–
St. Helena	4	5	40%	44%	–	60	–	–
St. Kitts and Nevis	52	61	32%	38%	–	135	–	–
St. Lucia	174	198	28%	33%	100%	158	11,483	1.3
St. Pierre and Miquelon	6	6	91%	92%	–	120	–	–
St. Vincent and the Grenadines	109	111	49%	58%	100%	84	32,320	< 0.5
Samoa	179	188	20%	22%	100%	403	109,932	7
São Tomé and Príncipe	165	216	62%	72%	100%	209	142,563	–
Saudi Arabia	26,246	34,176	82%	85%	30%	2,640	–	292
Senegal	12,861	17,861	42%	49%	83%	531	147,221	1,853.00
Serbia	9,856	9,720	56%	62%	8%	0	–	–
Seychelles	85	91	55%	64%	100%	491	1,288,643	29

7 **Marine Protected areas** number *2008*	8 **Fisheries capture** Tons of fish, shellfish, and molluscs captured *2008*	9 **Aquaculture** Tons of fish, shellfish and molluscs produced *2008*	10 **Fish species threatened, vulnerable or rare** number *2008*	11 **Improved sanitation** % of population with access *2008*	12 **Organic water pollutant (BOD) emissions** kg per day *2005*	
2	900	11	21	92%	–	Montenegro
–	50	-	–	96%	–	Montserrat
11	995,773	1,399	31	69%	72,779	Morocco
3	119,645	692	45	17%	–	Mozambique
4	372,822	28	21	33%		Namibia
–	39	-	–	–	–	Nauru
0	21,500	27,250	0	31%	-	Nepal
6	416,748	46,622	11	100%	122,052	Netherlands
7	16,698	0	15	–	–	Netherlands Antilles
21	3,719	2,108	17	–	–	New Caledonia
87	451,052	112,358	14	–	64,193	New Zealand
5	29,810	16,078	21	52%	–	Nicaragua
0	29,960	40	2	9%	–	Niger
0	541,368	143,207	21	32%	–	Nigeria
–	200	-	–	100%	–	Niue
2	292	-	9	–	–	Northern Mariana Islands
17	2,430,842	843,730	14	100%	-	Norway
3	145,631	120	20	–	6,498	Oman
5	451,414	135,098	22	45%	–	Pakistan
15	1,007	20	12	–	–	Palau
–	2,843	-	1	89%	–	Palestinian Territories
21	222,508	8,224	19	69%	13,719	Panama
24	223,631	92	38	45%	–	Papua New Guinea
0	20,000	2,100	0	70%	-	Paraguay
2	7,362,907	43,103	10	68%	–	Peru
212	2,561,192	741,142	60	76%	97,900	Philippines
3	142,496	36,813	6	90%	364,549	Poland
27	240,192	6,458	38	100%	105,041	Portugal
19	1,793	44	13	–	–	Puerto Rico
2	17,688	36	7	100%	3,328	Qatar
–	2,905	0	–	–	–	Réunion
10	5,410	12,532	16	72%	235,124	Romania
27	3,383,724	115,420	32	87%	1,425,913	Russian Federation
0	9,050	388	9	54%	-	Rwanda
–	794	-	–	–	–	St. Helena
1	450	0	14	96%	–	St. Kitts and Nevis
29	1,713	0	15	–	–	St. Lucia
–	–	–	–	–	–	St. Pierre and Miquelon
19	3,828	-	16	–	–	St. Vincent and the Grenadines
8	3,800	3	8	100%	–	Samoa
–	4,250	-	8	26%	–	São Tomé and Príncipe
3	68,000	22,253	16	–	-	Saudi Arabia
11	447,754	200	28	51%	-	Senegal
0	3,197	7,532	8	92%	–	Serbia
10	69,172	289	14	–	–	Seychelles

	1 Total population 1,000s		2 Urban population as % of total		3 Population within 100 km of coast as % of total population	4 Length of coastline km	5 Area of claimed EEZ km²	6 Mangrove forest area km²
	2010	*projected 2025*	*2010*	*projected 2025*	*2000*	*2010*	*2000*	*1997*
Sierra Leone	5,836	8,112	38%	46%	55%	402	–	1,838.00
Singapore	4,837	5,362	100%	100%	100%	193	–	6
Slovakia	5,412	5,413	55%	57%	0%	0	–	–
Slovenia	2,025	2,050	50%	52%	61%	47	–	–
Solomon Islands	536	725	19%	26%	100%	5,313	1,377,128	642
Somalia	9,359	13,922	37%	46%	55%	3,025	–	910
South Africa	50,492	53,766	62%	69%	39%	2,798	–	11
Spain	45,317	49,265	77%	81%	68%	4,964	683,236	–
Sri Lanka	20,410	22,033	14%	17%	100%	1,340	500,750	89
Sudan	43,192	56,688	40%	51%	3%	853	–	937
Suriname	524	586	69%	75%	87%	386	119,050	1,150.00
Swaziland	1,202	1,455	21%	24%	31%	0	–	–
Sweden	9,293	9,915	85%	87%	88%	3,218	73,166	–
Switzerland	7,595	8,020	74%	76%	0%	0	–	–
Syrian Arab Republic	22,505	28,592	56%	63%	35%	193	–	–
Taiwan	22,900	23,100	–	–	100%	1,566	–	–
Tajikistan	7,075	9,075	26%	30%	0%	0	–	–
Tanzania	45,040	67,394	26%	35%	21%	1,424	204,294	1,155.00
Thailand	68,139	72,628	34%	42%	39%	3,219	176,540	2,641.00
Togo	6,780	9,282	43%	54%	45%	56	10,807	26
Tokelau	1	2	0%	0%	–	101	–	–
Tonga	104	112	23%	28%	98%	419	844,978	10
Trinidad and Tobago	1,344	1,388	14%	21%	100%	362	60,659	< 70.0
Tunisia	10,374	11,797	67%	73%	84%	1,148	–	–
Turkey	75,705	87,364	70%	76%	58%	7,200	176,643	–
Turkmenistan	5,177	6,072	50%	57%	8%	0	–	–
Turks and Caicos Islands	33	38	93%	97%	–	389	–	–
Tuvalu	10	11	50%	58%	–	24	–	–
Uganda	33,796	53,406	13%	18%	0%	0	–	–
Ukraine	45,433	41,617	69%	74%	21%	2,782	86,392	–
United Arab Emirates	4,707	6,109	84%	88%	85%	1,318	21,200	30
United Kingdom	61,899	66,601	80%	83%	99%	12,429	–	–
United States of America	317,641	358,735	82%	86%	43%	19,924	8,078,169	1,990.00
Uruguay	3,372	3,546	92%	94%	79%	660	110,500	–
Uzbekistan	27,794	32,715	36%	40%	3%	0	–	–
Vanuatu	246	338	26%	34%	100%	2,528	530,162	16
Venezuela	29,044	35,370	93%	95%	73%	2,800	385,674	2,500.00
Vietnam	89,029	102,054	30%	41%	83%	3,444	237,800	2,525.00
Virgin Islands (UK)	23	26	41%	48%	100%	80	–	5.9
Virgin Islands (US)	109	103	95%	97%	100%	188	–	6
Wallis and Futuna Islands	15	17	0%	0%	–	129	–	–
Western Sahara	530	775	82%	85%	–	1,110	–	–
Yemen	24,256	35,509	32%	42%	64%	1,906	464,966	81
Zambia	13,257	18,890	36%	41%	0%	0	–	–
Zimbabwe	12,644	16,780	38%	47%	0%	0	–	–

7 **Marine Protected areas** number *2008*	8 **Fisheries capture** Tons of fish, shellfish, and molluscs captured *2008*	9 **Aquaculture** Tons of fish, shellfish and molluscs produced *2008*	10 **Fish species threatened, vulnerable or rare** number *2008*	11 **Improved sanitation** % of population with access *2008*	12 **Organic water pollutant (BOD) emissions** kg per day *2005*	
0	203,582	0	16	13%	–	Sierra Leone
3	1,623	3,518	22	100%	34,458	Singapore
0	1,655	1,071	7	100%	51,428	Slovakia
3	869	1,315	24	100%	28,767	Slovenia
21	26,235	1	12	–	–	Solomon Islands
2	30,000	-	26	23%	–	Somalia
30	643,686	3,215	65	77%	191,929	South Africa
47	917,188	249,062	52	100%	379,728	Spain
14	327,575	7,474	31	91%	-	Sri Lanka
1	65,500	2,000	13	34%	-	Sudan
7	23,811	38	20	84%	–	Suriname
0	70	0	3	55%	–	Swaziland
477	231,336	7,595	12	100%	97,622	Sweden
0	1,582	1,214	11	100%	–	Switzerland
4	6,996	8,595	27	96%	-	Syrian Arab Republic
–	1,016,390	323,982	–	–	–	Taiwan
0	146	26	8	94%	-	Tajikistan
17	325,476	15	138	24%	–	Tanzania
19	2,457,184	1,374,024	50	96%	-	Thailand
1	20,000	126	16	12%	–	Togo
–	200	-	–	93%	–	Tokelau
11	2,141	1	9	96%	-	Tonga
13	13,833	0	19	92%	-	Trinidad and Tobago
4	100,241	3,328	20	85%	–	Tunisia
13	494,124	152,260	60	90%	-	Turkey
0	15,000	16	12	98%	–	Turkmenistan
–	6,133	0	–	–	–	Turks and Caicos Islands
–	2,200	0	–	84%	–	Tuvalu
0	450,000	52,250	54	48%	-	Uganda
15	195,449	15,400	20	95%	527,203	Ukraine
3	74,075	1,206	9	97%	–	United Arab Emirates
149	596,004	179,187	34	100%	521,716	United Kingdom
787	4,349,853	500,114	164	100%	1,889,365	United States of America
4	110,691	36	28	100%	–	Uruguay
0	2,800	3,418	8	100%	–	Uzbekistan
21	60,881	40	11	52%	–	Vanuatu
19	295,364	18,627	29	–	–	Venezuela
36	2,087,500	2,461,700	33	75%	470,233	Vietnam
–	1,200	10	–	100%	–	Virgin Islands (UK)
15	1,065	-	11	–	–	Virgin Islands (US)
–	600	-	–	96%	–	Wallis and Futuna Islands
–	–	-	–	–	–	Western Sahara
1	127,132	5,640	18	52%	1,282	Yemen
0	79,403	-	10	49%	–	Zambia
0	10,500	2,450	3	44%	-	Zimbabwe

SOURCES

GLOSSARY

Definitions for technical terms were derived from:
Biodiversity and conservation glossary, The Marine Life Information Network: www.marlin.ac.uk
CORIS, The National Oceanic and Atmospheric Administration's coral reef information system: www.nos.noaa.gov/coris_glossary
FAO Fisheries Glossary: www.fao.org/fi/glossary
Intergovernmental Panel of Climate Change glossary: www.ipcc.ch/pub/gloss.pdf.
US Geological Survey definitions: http://toxics.usgs.gov/definitions

22–23 PART 1: PEOPLE AND COASTS

24–25 POPULATION GROWTH ALONG COASTS

Creel L. *Ripple effects: population and coastal regions.* Washington DC: Population Reference Bureau; 2003 September. p.1–8.
Crosset M, et al. *Population trends along the coastal United States, 1980-2008.* Washington DC: NOAA; 2008. p.1–2.
De Souza M. *Harmonizing population and coastal resources in the Philippines.*
Washington DC: Worldwatch Institute; 2004 August 15. p.1–2.
Hinrichsen D. *Coastal waters of the world: trends, threats and strategies.* Washington DC: Island Press; 1998. p.7–16.
WRI. *Populations losing livelihoods to polluted water.* Washington DC: WRI; 2008 April 2. p.1–2.

POPULATION OF US COASTAL COUNTIES
Crosset M, et al. op. cit.

COASTAL POPULATIONS
WRI. Coastal & marine ecosystems – fisheries: population within 100 km of coast; 2000. www.wri.org

POLLUTED OCEANS
UNEP/UN-HABITAT. *Sick water?* 2010. www.unwater.org

23 OF THE 25 MOST DENSELY POPULATED...
Crosset M, et al. pp.1–2. op. cit.

IN SOUTH ASIA, 825 MILLION...
WRI. *Populations losing livelihoods to polluted water.* op. cit.

26–27 URBANIZED COASTLINES

Dronkers J. *Pressures, impacts and policy responses in European coastal zones.* SPICOSA (accessed 2010 September 13). www.coastalwiki.org./spicosa
Neale D, Mohammed A. The urbanisation of Caribbean coastlines: a case study of the Trinidad west coast. Royal Institution of Chartered Surveyors (accessed 2010 September 13). www.rics.org
Torrey BB. Urbanization: an environmental force to be reckoned with. Population Reference Bureau (accessed 2010 September 13). www.prb.org
UNFPA. *State of world population 2007.* New York: UNFPA; 2007. p.1–99.
UN Population Division. World urbanization prospects, the 2009 revision. New York: UN. 2009. http://esa.un.org

GROWING URBANIZATION
UN. Proportion urban population from World urbanization prospects, the 2009 revision (accessed 2010 September13). http://esa.un.org

URBAN WORLD
UN. Proportion urban population. op. cit.

THE NUMBER OF MEGACITIES...
UN. World urbanization prospects, the 2009 revision. op. cit.

CITY GROWTH
UN. Thirty largest cities from World urbanisation prospects, the 2009 revision. op. cit.

72% OF SUB-SAHARAN AFRICA...
UNHabitat. State of the World's Cities: Trends in sub-Saharan Africa (accessed 2010 September 13). www.unhabitat.org

28–29 SUB-SAHARAN AFRICA

UNFPA. *State of world population 2009.* New York: UNFPA; 2009. p.91.
Wetlands International. www.ramsar.org
WCMC/UNEP. Mangroves of East Africa. UNEP; 2003.
World Bank.Valuing coastal and marine ecosystem services. Washington DC: World Bank; 2009. pp.24–27. www.siteresources.worldbank.org

DAR ES SALAAM IS GROWING...
UN. World population prospects, the 2009 revision population database. NY: UN; 2010.

URBAN POPULATION GROWTH
UNFPA, *State of world population 2009;* p.91. op. cit.

MAJOR COASTAL CITIES
UN. World population prospects, the 2009 revision population database. NY: UN; 2010.

FISHERIES BUSINESS
World Bank. pp.24–27. op.cit.

AFRICA'S COASTS AND OCEANS
World Bank. pp.24–27. op.cit.

MANGROVE DECLINE
WCMC/UNEP. Mangroves of East Africa op. cit. p.17.

30–31 ERODING SHORELINES

Alaska: portions of Arctic coastline eroding, no end in sight, says new study by University of Colorado, Boulder, *Space & Earth.* 2009 December 14 (reported by Physorg.com).

Dept of Labor and Natural Resources. Hawaii: coastal erosion and beach loss in Hawaii. Honolulu, Hawaii; 2000.

NOAA. Coastal management, who moved the beach? *Oceanservices*; 2010 May 4.

Stewart R. Our ocean planet: oceanography in the 21st century, an online textbook, p.1–4. http://oceanworld.tamu.edu/

STATE OF EUROPE'S COASTS

Niesing H. *Coastal erosion in Europe: A need for action. Results and proposals from the Eurosion project.* www.coastms.co.uk

EUROPE'S ERODING COASTLINE

European Commission. *Living with coastal erosion in Europe: sand & space for sustainability.* www.eurosion.org

15km² OF COASTAL LAND...

European Commission puts spotlight on coastal erosion. Brussels: EC; 2004 May 17.

ANNUAL US COASTAL CHANGE

National Assessment Synthesis Team, US Global Change Research Program. Climate change impacts on the United States: The potential consequences of climate variability and change; 2000. www.usgcrp.gov

66% OF CALIFORNIA'S BEACHES ARE ERODING...

NOAA. op. cit.

BIG SUR...

Hapke CH, Green KR. Rates of landsliding and cliff retreat along the Big Sur coast, California – measuring a crucial baseline. USGS Fact Sheet 2004–3099. www.pubs.usgs.gov

32–33 PART 2: MAJOR THREATS TO OCEAN RESOURCES

34–35 MARINE ECOSYSTEMS UNDER THREAT

Halpern BS, Walbridge S, Selkoe K.A, et al. A global map of human impact on marine ecosystems. *Science*; 2008. 319: 948–952.

JELLYFISH SWARMS...

Special Report: Jellyfish gone wild. National Science Foundation, 2008 November. www.nsf.gov

36–37 OCEAN DEAD ZONES

Diaz RJ, Rosenberg R. Spreading Dead Zones and Consequences for Marine Ecosystems. *Science;* 2008 Aug 15; vol.32:926–929. www.sciencemag.org

Goddard Earth Sciences Data & Information Services Center. Science focus: dead zones. http://disc.sci.gsfc.nasa.gov

Vaquer-Sunyer R, Duarte CM. Thresholds of hypoxia for marine biodiversity. *PNAS*. 2008 Oct 7; 105(40):15452-57. www.ncbi.nlm.nih.gov

LOSS OF POTENTIAL FISH FOOD

Diaz RJ, Rosenberg R. op.cit.

THE GULF OF MEXICO

Bruckner M. The Gulf of Mexico dead zone. Microbial life – educational resources. serc.carleton.edu

HYPOXIC & EUTROPHIC COASTAL AREAS

WRI. World hypoxic and eutrophic coastal areas. www.wri.org

GLOBALLY, COASTAL DEAD ZONES...

Diaz RJ, Rosenberg R. op. cit.

38–39 OCEAN DEAD ZONES: THE BALTIC SEA UNDER THREAT

Helsinki Commission. www.helcom.fi

Hinrichsen D. *Coastal waters of the world: trends, threats and strategies.* Washington DC: Island Press; 1998. p.51.

LEAD POLLUTION

HELCOM. Waterborne inputs of heavy metals to the Baltic Sea. www.helcom.fi

HELSINKI CONVENTION & NITROGEN & PHOSPHORUS

HELCOM. The Baltic Sea joint comprehensive environmental action plan. Helsinki: Helsinki Commission; 1993. pp.1–5.

HELCOM. Waterborne inputs of heavy metals to the Baltic Sea. www.helcom.fi

PRESSURES ON THE BALTIC

Pawlak J, Laamanen M, Anderson J. *Eutrophication in the Baltic Sea.* HELCOM; 2009.

PROBLEMS IN THE BALTIC

HELCOM. The Baltic Sea joint comprehensive environmental action plan. op. cit.

FISHERIES

HELCOM. List of threatened and declining species of lampreys and fishes of the Baltic Sea. Baltic Sea Environment Proceedings; 2007. p.109.

EFFECT OF POPs ON BIRD HEALTH

Helander B, Bignert A. Predatory bird health – white tailed sea eagle. HELCOM. www.helcom.fi

ALGAL BLOOM

Erkman I. Algae brings despair to Sweden. *New York Times*. 2005 Sept 12.

MERCURY

HELCOM. Waterborne inputs of heavy metals to the Baltic Sea. op. cit.

40–41 OCEAN DEAD ZONES: THE NORTHWEST PACIFIC

GIWA. Challenges to international waters: regional assessments in a global perspective. GIWA final report; 2006. pp. 38–53. www.unep.org/dewa/giwa

LOICZ, Draft executive summary of workshop held in February 2001, the East Asian Basins workshop.

Northwest Pacific Action Plan, Pollution Monitoring Regional Activity Centre. *State of the marine environment in the NOWPAP region*; 2007.

http://dinrac.nowpap.org
WWF. *Breathless coastal seas*. WWF Briefing Paper 2008, pp.3–19.

80% OF MARINE POLLUTION...

State of the marine environment in the NOWPAP region. op. cit.

MAJOR ACTIVITIES...

LOICZ. op. cit.

HARMFUL ALGAL BLOOM EVENTS

State of the marine environment in the NOWPAP region. p. 33. op. cit.

DYING WATERS

Eutrophication hotspots: *Breathless coastal seas*. op. cit.
Harmful algal blooms: *State of the marine environment in the NOWPAP region*. op. cit.

POLLUTED RIVERS

State of the marine environment in the NOWPAP region. p. 30. op. cit.

42–43 KEY COASTAL ENVIRONMENTS AT RISK

Adeel Z, King C. editors. *Conserving our coastal environment*. United Nations University, Japan; 2002. pp. 5–35.
Pope F. Marine plant life holds the secret to preventing global warming. London: *The Times*. 2009 Oct 14.
UNEP. East Asian Seas Region. Bangkok: UNEP; 2005. www.unep.org
UNEP, FAO, IUCN and CSIC. *Blue carbon, the role of healthy oceans in binding carbon*; 2009. www.grida.no
USGS. National Wetlands Research Center. Without restoration, coastal land loss to continue. 2003 May.

25% Of Salt Marshes...

Blue Carbon, the role of healthy oceans in binding carbon. op. cit.

COASTAL ECOSYSTEMS

IUCN. Global overview of wetland and Marine Protected Areas on the World Heritage list. www.unep.wcmc.org
World Bank. *World Development Report 2010*. The World Bank: Washington DC; 2010.

VALUABLE ECOSYSTEMS

R. Costanza, et al. The value of the world's ecosystem services and natural capital. *Nature* 387; 1997 May 15: 253–260.

BLUE CARBON SINKS

Blue carbon, the role of healthy oceans in binding carbon. op. cit.

44–45 ENVIRONMENTS AT RISK: SEAGRASSES

J. Loney. Loss of world's seagrass beds seen accelerating. Reuters Planet Ark; 2009 July 3. pp. 1–2.
UNEP/WCMC. *World atlas of seagrasses*. UNEP/WCMC: University of California Press: 2003. pp. 5–286.
UNU. Conserving our coastal environment, A summary of UNU's research on sustainable management of the coastal hydrosphere in the Asia Pacific Region. United Nations University; 2002. pp. 5–39.

SEAGRASSES OVER TIME

Waycott M, et al. Accelerating loss of seagrasses across the globe threatens coastal ecosystems. *PNAS*; 2009 July 28: 13; 3; 12377–12381. www.pnas.org

RISING RATE OF CHANGE

Waycott M, et al. op. cit.

35% OF DOCUMENTED...

Waycott M, et al. op. cit.

SEAGRASS MONITORING

Waycott M, et al. op. cit.

GLOBALLY THE WORLD LOSES...

J. Loney. op. cit.

DISTRIBUTION AND DIVERSITY OF SEAGRASSES

Global Seagrass distribution from version 2.0 of the global polygon and point dataset compiled by UNEP World Conservation Monitoring Centre (UNEP-WCMC), 2005.

AUSTRALIA HAS THE MOST DIVERSITY...

UNEP/WCMC. *World Atlas of Seagrasses*. op. cit.

46–47 ENVIRONMENTS AT RISK: MANGROVES

FAO, *The world's nangroves 1980-2005*. FAO Forestry Paper No. 153, FAO: Rome; 2007.
Kathiresan K. *Mangrove ecosystems, distribution of mangroves*. FAO; 2003. pp. 92–100.
ITTO. *Mangroves: forests worth their salt*. ITTO: Japan. 2003. pp.1–7.
Spalding M, Kainuma M, Collins L. *World atlas of mangroves*. Earthscan: London and Washington; 2010.
Valiela I, Bowen J, York J. Mangrove forests: one of the world's threatened major tropical environments, *BioScience*. 2001 October: 51;10;807–814.

THERE WERE 18.8 MILLION...

FAO. *The world's mangroves, 1980–2005*. op. cit.

MANGROVES

Wet Tropics Management Authority. Plant diversity – mangroves. www.wettropics.gov.au

DISTRIBUTION AND DIVERSITY OF MANGROVES

Mangroves extracted from version 3.0 of the global polygon dataset compiled by UNEP World Conservation Monitoring Centre (UNEP-WCMC) in collaboration with the International Society for Mangrove Ecosystems (ISME), 1997.

THE LONGTERM SEQUESTRATION...

Pidgeon E. Carbon sequestration by coastal marine habitats: important missing sinks. In Laffoley D, Grimsditch G, editors. *The management of coastal carbon sinks*. IUCN; 2009 November.

FISH FROM TREES

Mangroves: forests worth their salt. op. cit.

MANGROVE DECLINE

WCMC/UNEP. Mangroves of East Africa. UNEP; 2003. p.17.

RETAINING MANGROVES

Gilman EL, et.al. Threats to mangroves from climate change and adaptation options. *Aquatic Botany.* AQBOT 2097; 2008.

48–49 CORAL REEFS IN DANGER

Donner S, Potere D. The inequity of the global threat to coral reefs. *BioScience*; 2007 March: vol 57; no3; 214–215

Eakin CM. Impacts of ocean acidification on coral reefs, NOAA Coral Reef Watch; 2007. Power point slides 1–21.

IUCN. Ocean acidification, seas turning sour, briefing paper. 2008 December. pp.1–2. www.iucn.org

Laffoley D, Baxter JM, editors. Ocean acidification: A special introductory guide for policy advisers and decision makers. Ocean Acidification Reference User Group.

European Project on Ocean Acidification; 2009. pp.1–12.

Natural Resources Defense Council. Ocean acidification: the other CO_2 problem; 2009. pp.1-2. www.nrdc.org

Pew Center on Global Climate Change. The science and consequence of ocean acidification. *Science Brief No. 3;* 2009 Aug. pp.1–8.

Wilkinson C, editor. *Status of coral reefs of the world: 2008.* Townsville, Australia: Global coral reef monitoring network & reef and rainforest research centre; 2008. www.gcrmn.org

Wilkinson C, editor. *Status of coral reefs of the world: 2008, Executive Summary.* UNEP, NOAA, IUCN, UNESCO, Great Barrier Reef Marine Park Authority; 2008. pp.1–4.

WRI. Ocean acidification, the other threat of rising CO_2 emission. *Earthtrends*; 2007 Sept. pp.1–6. www.earthtrends.wri.org

THE MASTER BUILDERS OF THE SEA

Hinrichsen D. Requiem for Reefs. *International Wildlife*; 1997 March/April. p. 14–15.

THE STATE OF CORAL

Coral reefs from version 7.0 of the global 1 km raster dataset compiled by the UNEP World Conservation Monitoring Centre (UNEP-WCMC); 2003.

Wilkinson C. p.11. op. cit.

CARIBBEAN

Wilkinson C. p.17. op. cit.

AUSTRALIA

Hinrichsen D. Reefs at Risk. *Defenders*; Summer 1999. p.7–10.

CORAL BLEACHING

International Coral Reef Initiative. Status and threat to coral reefs. www.iyor.org

THE FUTURE FOR CORAL

Wilkinson C. op. cit.

50–51 THE EMPTY OCEAN

Stokstad E. Ecology: Global loss of biodiversity harming ocean bounty. *Science* 2006 Nov; 314(5800):745.

Worm B, Barbier EB, Beaumont NJ, et al. Impacts of biodiversity loss on ocean ecosystem services. *Science;* 2006 Nov; 314(5800):787–790

STATE OF TUNA

Majkowski J. Tuna and tuna-like species: global status of fishery resources. FAO; 2007.

FISH PRODUCTION

FAO. *World review of fisheries and aquaculture 2008.* pp. 31–32. www.fao.org

FAO Yearbooks of Fishery Statistics, Summary tables: World fisheries production by capture & aquaculture by country 2008. www.fao.org/fishery

STATE OF FISH STOCKS

FAO. *World review of fisheries and aquaculture 2008.* p.30.

NEARLY 80% OF FISH...

FAO. *State of the world's fisheries 2008.* Figure 21. p.33

LOSS OF STOCKS

Black R. 'Only 50 years left' for sea fish. BBC; 2006 Nov 2.

UNLESS WE FUNDAMENTALLY CHANGE...

Black R. op.cit.

52–53 PART 3: TRADE, COMMERCE AND TOURISM

54–55 MAJOR SHIPPING LANES

International Maritime Organisation. International Shipping: Carrier of World Trade. IMO: London; 2009.

Kaluza P, Kölzsch A, Gastner M, Bernd Blasius, J. The complex network of global cargo ship movements. Journal of the *Royal Society: Interface*; 2010; 7: 1093–1103.

UNEP. Global marine oil pollution information gateway. http://oils.gpa.unep.org/facts/sources.htm

Vidal J. Health risks of shipping pollution have been underestimated. *The Guardian*; 2009 April 9.

GLOBAL SHIPPING ROUTES

Halpern BS, Walbridge S, Selkoe K.A, et al. A global map of human impact on marine ecosystems. *Science*; 2008; 319: 948–952.

85% OF ALL SHIPS... & SHIPPING IS RESPONSIBLE FOR...

Vidal J. op. cit.

WORLD'S TOP SHIPPING PORTS

American Association of Port Authorities. www.aapa-ports.org

CLIMATE CHANGES OPENS FABLED...

Weir F. Global warming opens new arctic shipping lane. *Christian Science Monitor.* 2009 October 15.

56–57 ENERGY FROM THE SEA: OIL AND GAS

Congressional Research Service. The Deepwater Horizon

oil spill: coastal wetlands and wildlife: impacts and response. CRS Report to Congress, 2010 August 5

Israel B. Degraded oil from BP spill coats seafloor. *Live Science*; 2010 Sept 21.

The BP spill: has the damage been exaggerated? *Time Magazine*; 2010 July 29.

UNEP. Global marine oil pollution information gateway, sources of oil to the sea. http://oils.gpa.unep.org.

Water Encyclopedia. Oil spills: impact on the ocean. www.waterencyclopedia.com

World Conservation Union. *Environmental management of offshore oil development and maritime oil transport.* IUCN Commission on Environment, Economic, and Social Policy; 2004 October. pp.1-50.

OFFSHORE OIL RESERVES

Chakhmakhchev A, Rushworth P. Global overview of offshore oil & gas operations for 2005-2009. *Offshore Magazine*; 2010 May 1. www.offshore-mag.com

The world offshore oil and gas production spend forecast 2009–2013. www.energy-market-research.info

Offshore oil and gas industry of Russia and CIS: outlook to 2020. PRLog. 2009 Feb 17.

The CIA World Factbook. www.cia.gov

REDUCED POLLUTION

Patin S. *Oil pollution of the seas.* www.offhore-environments.com

SOURCES OF POLLUTION

National Research Council of the US National Academy of Sciences. www.nationalacademies.org/nrc/

Patin S. *Oil pollution of the seas.* op. cit.

UNEP. Global marine oil pollution information gateway, quoting GESAMP figures (Joint Group of Experts on the Scientific Aspects of Marine Environmental Protection). http://oils.gpa.unep.org/facts/sources.htm

POTENTIAL NEW ENERGY SOURCE...

Pearce F. Ice on fire: the next fossil fuel? *The New Scientist*; 2009 June 24.

US Department of Energy. Methane hydrate: the gas resource of the future. www.energy.gov

58–59 ENERGY FROM THE SEA: WIND

Global Offshore Wind Farms Database. www.4coffshore.com/offshorewind/ (accessed 2010 October 23).

Offshore wind market seen doubling in 2010. Reuters PlanetArk; 2010 October 21.

Knowledge Transfer Network. Key statistics: offshore wind; 2009 December. http://ktn.innovateuk.org

DTI (UK). Sustainable energy technology route maps. Offshore wind energy. http://ti.gov.uk

Worldwatch Institute. Vital signs 2009. Wind power continues rapid rise. Worldwatch Institute: Washington DC; 2009.

OFFSHORE WIND FARMS IN OPERATION & IN THE FUTURE

4C Offshore online global wind farms database. www.4coffshore.com/offshorewind/

60–61 ENERGY FROM THE SEA: TIDES AND WAVES

Appleyeard D. Ocean energy developments. *Renewable Energy World Magazine*; 2009 September 18.

BBC. Global energy guide: wave and tidal power. www.bbc.co.uk (accessed 2010 October 21).

Bhuyan G. Harnessing the power of the oceans. International Energy Agency; *IEA Open Energy Technology Bulletin*. International Energy Agency; 2008 July; No. 52.

Johnson S. World first wave and tidal energy projects for Scotland. *Telegraph-Scotland*; 2010 March 17.

Maser M. Tidal energy: a primer. Blue Energy. Canada; 2009.

National Research Council Canada. The tide rolls in on ocean energy; 2010 October 4.

Renewable ocean energy: tides, currents and waves. *Alternative Energy News*. 2006 September 18.

THE POWER OF THE WAVES...

BBC. Global energy guide. op. cit.

UK ENERGY PROJECTS

Johnson S. op. cit.

TIDAL ENERGY

Khan J, Bhuyan G. Ocean energy: global technology development status. Report for IEA-OES. www.iea-oceans.org

Maser M. op. cit.

National Research Council Canada. op. cit.

Renewable ccean energy: tides, currents and waves. op. cit.

BBC. Global energy guide. op. cit.

DEVELOPMENT OF TIDAL ENERGY, DEVELOPMENT OF WAVE ENERGY, OCEAN BASED ENERGY SYSTEMS

Khan J, Bhuyan G. op. cit.

62–63 COASTAL AND MARINE TOURISM

Burke L. et.al. *Reefs at risk in the Caribbean: marine-based sources of threat.* WRI: Washington DC; 2004.

Caribbean Tourism Organization. Caribbean tourism, latest statistics, 2009. Caribbean Tourism Organization; 2010 August.

Honey M, Krantz D. *Global trends in coastal tourism.* For Marine Program WWF-US. Washington DC; 2007 December. www.responsibletravel.org

UN World Tourism Organization. *UNWTO Tourism highlights 2010 edition.* www.unwto.org/facts

UNEP/MAP-Plan Bleu. *State of the environment and development in the Mediterranean.* Athens; 2009.

MAJOR TOURIST DESTINATIONS

UN World Tourism Organization. op. cit.

TOURISM IS THE LARGEST SECTOR OF THE GLOBAL...

Honey M, Krantz D. op. cit.

INCREASING TOURISM

UN World Tourism Organization. op. cit.

IMPACT OF TOURISTS ON COASTAL ECOSYSTEMS

Orams M.B. *Biodiversity and tourism: conflicts on the world's seacoasts and strategies for their solution.* German Federal Agency for Nature Conservation: Bonn; 1997. pp. 51–53.

Honey M, Krantz D. op. cit.

64–65 COASTAL AND MARINE TOURISM: THE MEDITERRANEAN

De Stefano L. *Freshwater and tourism in the Medieterranean.* WWF Mediterranean Program: Rome; 2004 June. www.panda.org.mediterranean

UNEP/MAP-Plan Bleu. *State of the environment and development in the Mediterranean.* Athens; 2009.

WWF. Tourism Threats in the Mediterranean, background information. WWF. pp. 14 (accessed 2010 November 21). www.monarchus-guardian.org

WWF. Thirteen key Mediterranean areas in need of protection. WWF (accessed November 21). www.assets.panda.org

MEDITERRANEAN ENVIRONMENT

UNEP/MAP-Plan Bleu. op. cit.

THE MEDITERRANEAN IS THE WORLD'S LEADING...

De Stefano L. op. cit.

14% OF THE MEDITERRANEAN COAST...

WWF. Thirteen key Mediterranean areas in need of protection. op. cit.

THE DEMISE OF MEDITERRANEAN FISHERIES

UNEP/MAP-Plan Bleu. op. cit.

ENDANGERED MEDITERRANEAN SPECIES

WWF. Tourism threats in the Mediterranean, background information. op. cit.

WASTEWATER BURDENS

UNEP/MAP-Plan Bleu. op. cit.

TOURISM AND RECREATION...

UNEP/MAP-Plan Bleu. op. cit.

66–67 FARMING THE SEA

Environmental Justice Foundation. *Farming the sea, costing the earth.* Environmental Justice Foundation: London; 2004.

International Food Policy Research Institute, World Fish Center, FAO. *Outlook for fish to 2020 – meeting global demand.* IFPRI-FAO. World Fish Center; 2003 October. pp. 1-23.

Mock G, White R, Wagener A. Farming fish: the aquaculture boom. *Earthtrends*; updated 2001 July. http://earthtrends.wri.org

Naylor R, et. al. Effect of aquaculture on world fish supplies. *Nature*; 2000 June 29; vol.405. pp. 1017–1024.

INCREASING IMPORTANCE

FAO. *State of the world fisheries and aquaculture 2008.* FAO: Rome; 2009. www.fao.org

WORLD MARICULTURE INCREASE

FAO. op. cit.

FARMED FISH, SHELLFISH AND AQUATIC PLANTS

FAO. op. cit.

5 KG OF WILD FISH

Naylor R, et al. op. cit.

68–69 FARMING THE SEAS: ASIA AND THE INDO-PACIFIC REGION

Lovatelli A, editor. *The future for mariculture: a regional approach for responsible development in the Asia-Pacific region.* FAO/NACA Regional Workshop: Guangszhou, China; 2006 March 7–11.

Primavera JH. Overcoming the impacts of aquaculture on the coastal zone. *Ocean and Coastal Management*; 49; 2006. pp. 531–545. www.sciencedirect.com

Ramos MH, Rueca LM. Ecosystem approach to fisheries and aquaculture in the Philippines. PPP for Republic of Philippines Department of Agriculture; Bureau of Fisheries; Aquatic Resources Regional Office no. 4-A.

Ricohermso MA, Bueno PB, Sulit VT. Maximizing opportunities in seaweeds farming. http://news.seafdec.org

THE ASIA PACIFIC REGION IS REPONSIBLE FOR 91%...

Lovatelli A, editor. op. cit.

SEAWEED FARMING IN THE PHILIPPINES

Lovatelli A, editor. op. cit.

AQUACULTURE SECTORS

Lovatelli A, editor. op. cit.

EMPLOYMENT IN AQUACULTURE

FAO. *State of the world fisheries and aquaculture 2008.* FAO: Rome; 2009. www.fao.org

SMALL-SCALE AQUACULTURE

Lovatelli A, editor. op. cit.

ECONOMIC IMPORTANCE OF AQUACULTURE

Contributions of fisheries and aquaculture in Asia and the Pacific region. Status and potential of fisheries and aquaculture in Asia and the Pacific 2008. FAO, Regional Office for Asia and the Pacific (accessed 2010 November 23). www.fao.org

80% AQUACULTURE COMES FROM SMALL-SCALE...

Subasinghe R, Phillips MJ. Pro-poor seafood trade: challenges and investment opportunities for small scale aquaculture farmers. PPT at www.ifad.org

70–71 PART 4: CLIMATE CHANGE

72–73 THE OCEAN CONVEYOR BELT

Kleinen T, Osborn T. The thermohaline circulation, climatic research unit information sheet no. 7. www.cru.uea.ac.uk

Rahmstorf, S. The thermohaline ocean circulation, a brief fact sheet; Home page Stefan Rhamstorf, Potsdam Institute for Climate Impact Research. 26. August 2010. www.pik-potsdam.de

Environmental Literacy Council. The great ocean conveyor belt. www.enviroliteracy.org

THE GULF STREAM TRANSPORTS…

Cunninhgam S, Bryden H. The Atlantic conveyor belt. *Planet Earth*; 2004 Spring. www.nerc.ac.uk

SOME DEEP CURRENTS TAKE…

Surface and subsurface ocean currents. www.physicalgeography.net

THE GULF STREAM

European climate could change rapidly, over decades rather than centuries. Natural Environment Research Council. www.nerc.ac.uk

UPWELLINGS

Upwellings and downwellings. www.redmap.org,au

74–75 RISING SEAS

Barnes S. Study: Earth's polar ice sheets vulnerable to even moderate global warming. News at Princeton; 2009 December 16. www.princeton.edu

Church J, White N. A 20th century acceleration in global sea level rise. *Geophysical Research Letters*; 33, L01602. www.agu.org

CSIRO. Sea level rise. www.cmar.csiro.au

Global warming basics. Pew Center on Global Climate Change. www.pewclimate.org

Climate Institute. www.climate.org

BANGLADESH SEA LEVEL RISE

Impact of sea level rise in Bangladesh. Maps and graphics at UNEP-GRID-Arendal. www.maps.grida.no

NILE DELTA SEA LEVEL RISE

UNEP vital climate graphics. www.grida.no/climate/vital

SEA LEVEL RISE

CSIRO. Global mean sea level. www.cmar.csiro.au

EFFECT OF SEA LEVEL RISE

DIVA run by University of Southampton, courtesy Nicholls and Brown, using 2007 report of IPCC figures.

85% OF GLACIERS…

Cook A, et al. Retreating glacier fronts on the Antarctic Peninsula over the past half-century. *Science*; 2005 April 22. 308:572;541–544.

CÔTE D'IVOIRE… & KENYA…

UN Habitat, *State of the world's cities, 2008–2009*. UN Habitat: Nairobi; 2008.

Cocks T. Threat of rising seas looms over coastal Africa. Reuters; 2009 December 18.

76–77 RISING SEAS: SMALL ISLAND DEVELOPING STATES

Brown L. Rising sea level forcing evacuation of island country. Worldwatch Feature, Worldwatch Institute: Washington DC; 2001 November. pp. 1–3.

IPCC Working Group II: Impacts, adaptation and vulnerability. human systems: threatened small island states. IPCC. www.ipcc.ch

O'Reilly M. Defining environmental migrants. Worldwatch Institute: Washington DC; 2010 February 11.

UNEP. Maldives: State of the environment 2002. UNEP: Bangkok. 2003. www.rrcap.unep.org

VULNERABLE ISLANDS

Environmental Vulnerability Index. www.vulnerabilityindex.net/

TUVALU

Ralston H, Horstmann B, Holl C. Climate change challenges Tuvalu. Germanwatch; 2004. www.germanwatch.org

FOE Australia. The Pacific nations and the impact of global warming. Chain Reaction # 91; 2004. www.foe.org.au

MALDIVES

UNEP. Maldives: State of the Environment 2002. op. cit.

INCREASING RELIANCE ON WATER DESALINATION

Ibrahim S, Bari M, Miles L. Water resources management in Maldives with an emphasis on desalination. Maldives Water and Sanitation Authority. www.pacificwater.org

THE 51 SIDS PRODUCE ONLY…

UN-OHRLLS. The impact of climate change on the development prospects of the least developed countries and small island developing states; 2009. www.unohrlls.org

KIRIBATI

Kirby A. Islands disappearing under rising seas. BBC; 1999 June 14.

30% OF KNOWN THREATENED PLANT SPECIES… & 23% OF SIDS' BIRD SPECIES…

IPCC Working Group II: Impacts, adaptation and vulnerability. op. cit.

78–79 EXTREME WEATHER EVENTS

EM-DAT. www.emdat.be/disaster-profiles

Harmeling, S. *Global climate risk index, 2010.* Germanwatch; 2009 December.

Kleiner K. Climate science in 2009. *Nature*; 2009 December 17. www.nature.com

McMullen C. ed. *Climate change 2009*, Science Compendium. UNEP; 2009.

NOAA. Worldwide weather and climate events. www.ncdc.noaa.gov

World Bank. *World development report 2010.* World Bank: Washington DC; 2010.

A GROWING PROBLEM

World Bank. *World development report 2010.* op. cit.

EL NIÑO

NOAA. El Niño theme page. www.pmel.noaa.gov

WHO IS MOST AT RISK?

Harmeling, S. op. cit.

GREENLAND

Sea ice at lowest level in 800 years near Greenland. Science Daily; 2009 July 2. www.sciencedaily.com

MEXICO

Ellingwood K. Mexico water shortage becomes crisis amid

drought. *LA Times*; 2009 September 07. http://articles.latimes.com

BANGLADESH

Downpour halts life. 2009 July 29. www.bangladesh2day.com

AUSTRALIA

South-east heatwave hottest in in 70 years. 2009 January 28. www.abc.net.au

80–81 OCEAN ACIDIFICATION

An international symposium: The ocean in a high CO_2 World. www.ocean-acidification.net/

IUCN. *Ocean acidification: sea turning sour.* IUCN: Gland, Switzerland; 2008 December. pp. 1–2.

NRDC. Ocean acidification: the other CO_2 problem. NRDC: New York; 2009 September 17. pp. 1–2.

Ocean Acidification Reference User Group (2009). *Ocean acidification: the facts.* A special introductory guide for policy advisers and decision makers. Laffoley Dd'A. and Baxter JM. editors. European Project on Ocean Acidification (EPOCA). http://cmsdata.iucn.org

WRI, Earthtrends Environmental Information. Ocean acidification, the other threat of rising CO_2 emissions. WRI: Washington DC; 2007 September. pp. 1–5.

PAST AND PROJECTED OCEAN Ph LEVELS

NRDC. op. cit.

CHANGING ACIDITY

Sabine CL, Feely RA, Gruber RM. et al. The oceanic sink for anthropeogenic CO_2. *Science;* 2004; 305(5682):367-371.

COLD WATER CORAL REEFS

UNEP/GRID-Arendal, Coldwater coral reefs, distribution, UNEP/GRID-Arendal Maps and Graphics Library. http://maps.grida.no

BY 2100 UP TO 70%

Ocean Acidification Reference User Group. op. cit.

COLD WATER CORALS AT RISK

Ocean Acidification Reference User Group. op. cit.

82–83 DISAPPEARING ARCTIC

Arctic ice loss: northwest passage now open, says space agency. AFP; 2007 September 14.

Ahlenius H. UNEP-GRID map of Arctic Sea ice. Updated 2007 December. http://maps.grida.no

Barnes S. Study: Earth's polar ice sheets vulnerable to even moderate global warming. News at Princeton; 2009 December 16. www.princeton.edu

Center for International Environmental Law, Climate Change and Arctic Impacts. Ciel (accessed on 2010 August 30). www.ciel.org

Devlin H. Greenland's ice sheet is melting faster than ever, data shows. *The Times*; 2009 November 13. www.thetimes.co.uk

Roach J. Greenland ice sheet is melting faster, study says. *National Geographic News*; 2006 August 10.

Zabarenko D. Scant Arctic ice could mean summer double whammy. Reuters PlanetArk; 2010 February 5. pp 1–2.

IMPACT ON ARCTIC WILDLIFE

Goldenberg S. Scientists investigate massive walrus haul-out in Alaska. *The Guardian*; 2010 September 13.

DECREASE OF ARCTIC SEA ICE

National Snow and Ice Data Center. http://nsidc.org

THE DECREASE OF ARTCTIC SEA ICE

Ahlenius H. The decrease of Arctic sea ice, minimum extent 1982 & 2007, & climate projections. UNEP/GRID-Arendal; 2007 December. http://maps.grida.no

IF ALL GREENLAND...

Gornitz V. Sea level rise, after the ice melted and today. NASA Science Briefs; 2007 January. www.giss.nasa.gov

IN AUGUST 2010 AN ICEBERG...

Dell'Amore C. Ice island breaks off Greenland; bigger than Manhattan. *National Geographic*; 2010 August 6. http://news.nationalgeopgraphic.com

84–85 ANTARCTIC: LOSING ICE COVER

Black R. Antarctic to feed major sea rise. BBC News; 2009 December 2.

Chestney N. East Antarctic ice began to melt faster in 2006 – study. Reuters News Alert; 2009 November 22. http://uk.reuters.com

Hogan J. Antarctic ice sheet is an awakened giant. *New Scientist*; 2005February 2.

Ice shelves disappearing on Antarctic Peninsula: glacier retreat and sea level rise are possible consequences. Science Daily. 2010 February 22. www.sciencedaily.com

New map to help calculate antarctic ice loss. Our Amazing Planet. 2010 July 22. (accessed 2010 August 29). http://ouramazingplanet.com

Vaughan G. Antarctic Peninsula: rapid warming. *British Antarctic Survey*, 2009.

IF THE ANTARCTIC ICE SHEETS MELTED...

Bentley R, Thomas RH. Velicogna I. Ice sheets, global outlook for ice and snow. UNEP. www.unep.org

LOSS OF ICE MASS

UNEP/GRID-Arendal, Mass balance of the West Antarctic Ice Sheet, UNEP/GRID-Arendal Maps and Graphics Library. http://maps.grida.no

THE ANTARCTIC CONTAINS 90%...

Antarctica. The Encyclopedia of Earth. www.eoearth.org

ADELIE PENGUINS

Global climate change, Antarctica and krill. Antarctic Krill Conservation Project. www.krillcount.org

Wolf S. Climate change threatens penguins. Action Bioscience. (accessed 2010 November 21) www.actionbioscience.org

WILKINS ICE SHELF

Antarctic Ice shelf disintegration underscores a warming world. 2008 March 25. National Snow and Ice Data Center. http://nsidc.org

86–87 PART 5: SEAS IN CONFLICT

88–89 CONTESTED ISLANDS

American University. Abu Musa: Island dispute between Iran and UAE. http://www1.american.edu

Asia News Network. Stakeholders in the sea, China and US arbitrators in sovereignty disputes; 2010 September 27. www.asianewsnet. net

Blatant World. Disputed territories in Asia. www.blatantworld.com

Mayes S. Whose Hans? Faulty theory? *Canadian Geographic Magazine*; 2005. www.canadiangeographic.ca

Carroll R, Weaver M. Argentina to state Falkland Islands case to UN chief. The Guardian; 2010 February 24. www.guardian.co.uk

Mapping the dispute over Falklands oil. Off-shore Technology.com; 2010 April 9. www.off-shoretechnology.com

Mellgard P. Trouble brews in the South China Sea. Current Conflicts; 2010 June 22. http://currentconflicts.foreignpolicyblogs.com

Paracel/Sprately Islands Blogspot, Paracel and Spratly Islands Forum, 2008. http://paracelspratlyislands.blogspot.com

Sutter R. CRS Report for Congress, East Asia – disputed islands/offshore claims issues for US Policy; 1992 July 28.

Wain B. All at sea over resources in East Asia. Yale Global; 2007August 14. http://yaleglobal.yale.edu

90–91 PIRACY: A RECENT GROWTH INDUSTRY

ASI Global Maritime Response; 2010. http://asiglobalresponse.com

Chalk P. Maritime piracy: reasons, dangers and solutions. Rand Corporation briefing paper presented to US Congressional Committee; 2009 February 4.

ICC. 2009 Worldwide piracy figures surpass 400. ICC Commercial Crimes Services: London; 2010 January 14.

Hanson S. *Combating maritime piracy.* Council on Foreign Relations; 2010 January 7.

Harper M. Piracy off the coast of Somalia has made many people very rich. BBC News; 2009 May 24.

International Chamber of Commerce. International Maritime Bureau (IMB) Piracy Reporting Centre. www.icc-ccs.org

BETWEEN 2003 AND 2008 THRE WERE…

ASI Global Maritime Response. op. cit.

GLOBAL PIRACY

ICC Commercial Crime Services. IMB Piracy Map 2009. www.icc-ccs.org

UNODC. Maritime piracy, in, *A transnational organized crime threat assessment.* www.unodc.org

2009: 153 VESSELS BOARDED

ASI Global Maritime Response. op. cit.

PIRACY OFF THE COAST OF SOMALIA

UNOSAT. op. cit.

REPORTED SOMALI PIRATE INCIDENTS

UNOSAT. op. cit.

THE PRICE OF MODERN PIRACY

ASI Global Maritime Response. op. cit.

Harper M. op. cit.

92–93 PART 6: MANAGEMENT OF COASTAL AND MARINE AREAS

94–95 INTEGRATED COASTAL AND OCEAN MANAGEMENT

Burbridge P, Glavovic B, and Olsen S. *Practitioner reflections on ICM experience in Europe, South Africa and Ecuador*; 2010. Manuscript to be published.

Sorensen J. *Baseline 2000 background report – the status of integrated coastal management as an international practice,* second iteration; 2002 August 26. Available at: Harbor and Coastal Center, University of Massachusetts. www.uhi.umb.edu

Sorensen J. National and International Efforts at Integrated Coastal Management: Definitions, Achievements, and Lessons. *Coastal Management*; vol. 25, 1997, pp. 3–41.

96–97 INTERNATIONAL MANAGEMENT PLANS

Brugere C. Can integrated coastal management solve agriculture-fisheries-aquaculture conflicts at the land-water interface? A Perspective from New Institutional Economics. FAO; 2006.

European Commission, Integrated coastal zone management. http://ec.europa.eu

Sorensen J. *Baseline 2000 background report – the status of integrated coastal management as an international practice,* second iteration; 2002 August 26. Available at: Harbor and Coastal Center, University of Massachusetts. www.uhi.umb.edu

UNEP Regional Seas Program. www.unep.org

Yu H, Bermas N. *Integrated coastal management: PEMSEA's practices and lessons learned.* United Nations Institute for Training and Research; 2004.

INTEGRATED COASTAL ZONE MANAGEMENT IN EUROPE…

European Commission. Integrated coastal zone management. http://ec.europa.eu

98–99 INTERNATIONAL MANAGEMENT PLANS: THE MEDITERRANEAN

UNEP/Plan Bleu. *State of the environment and development in the Mediterranean; 2009.* pp. 9–200.

UNEP. Mediterranean Action Plan. www.unepmap.org

Mifsud P. Mediterranean Action Plan: an introduction. UNEP Regional Seas Programme; 2006.

100–01 MARINE PROTECTED AREAS

Brown L. Expanding marine protected areas to restore fisheries. Earth Policy Institute News Brief; 2008 November 13.

Green Bush. *The Economist*; 2009 January 10. p.71.

MPA Global. A database of the world's marine protected areas. www.mpa.global.org

Pala C. Marine reserves proliferate worldwide. *Cosmos Magazine*; 2009 January 7.

WWF. *Marine protected areas – providing a future for fish and people.* WWF: Gland, Switzerland; 2004. pp.1–19.

OCEANS COVER 70%...

MPA Global. A database of the world's marine protected areas. op. cit.

MARINE PROTECTED AREAS

Map compiled by the UNEP World Conservation Monitoring Centre (UNEP-WCMC), 2010.

TURTLE CONSERVATION

Ausubel J, Crist D, Waggoner P. *First census of marine life 2010 – Highlights of a decade of discovery*. Census of Marine Life International Secretariat: Washington DC; 2010.

AUSTRALIA'S GREAT BARRIER REEF...

Reef Ed. www.reefed.edu.au

102–03 LARGE MARINE ECOSYSTEMS (LMES)

UNEP Regional Seas Program linked with Large Marine Ecosystems Assessment and Management. www.lme.noaa.gov

White water to blue water: a large marine ecosystem strategy for the assessment and management of the Caribbean sea large marine ecosystem. www.lme.noaa.gov

UNEP/MAP. Strategic partnership for the Mediterranean LME. www.unepmap.org

104–05 MARINE ECOSYSTEMS AND SPECIES IN PERIL

IUCN. Red list of threatened species. IUCN; 2008. www.iucn.org

TUNA

Gronewold N. Is the bluefin tuna an endangered species? Scientific American; 2009 October 14. www.scientificamerican.com

MANATEE

Manatee. National Geographic Animals (accessed 2010 December 3). animals.nationalgeographic.com

GREEN SEA TURTLE

Marine turtles: Three of the seven existing species of marine turtle are critically endangered. WWF (accessed 2010 December 3) www.worldwildlife.org

OCEAN POLLUTION

UNEP/UN-HABITAT. *Sick water?* 2010. www.unwater.org

POLAR BEAR

Saving the polar bear. The Center for Biological Diversity (accessed 2010 December 3). www.biologicaldiversity.org

SOFT CORAL

Donner S, Potere D. The inequity of the global threat to coral reefs. *BioScience*; 2007 March: vol 57; no3; 214–215

Wilkinson C, editor. *Status of coral reefs of the world: 2008.* Townsville, Australia: Global coral reef monitoring network & reef and rainforest research centre; 2008. www.gcrmn.org

SEAGRASSES

Waycott M, et al. Accelerating loss of seagrasses across the globe threatens coastal ecosystems. *PNAS*; 2009 July 28: 13; 3; 12377–12381. www.pnas.org

MANGROVES

Mangrove forests in worldwide decline. IUCN Red List (accessed 2010 April 9). www.iucn.org

ANCHOVIES

FAO. State of the World Fisheries. FAO; 2008.

SEA URCHIN

How will ecosystems be affected? (accessed 2010 December 3). www.ocean-acidification.net

TUVALU

Ralston H, Horstmann B, Holl C. Climate change challenges Tuvalu. Germanwatch; 2004. www.germanwatch.org

FOE Australia. The Pacific nations and the impact of global warming. Chain Reaction # 91; 2004. http://foe.org.au/

106–15 TABLES

Col 1. UN Population Division. World population prospects: the 2008 revision.

Col 2. UN Population Division. World urbanization prospects: the 2009 revision.

Col 3. CIA World Factbook (accessed 2010 December 3) www.cia.gov

Col 4. Earthtrends. WRI. www.wri.org and CIA World Factbook (accessed 2010 November 23). www.cia.gov

Col 5. Earthtrends. WRI. www.wri.org

Col 6. Earthtrends. WRI. www.wri.org

Col 7. World Development Indicators. The World Bank (accessed 2010 November 23). www.worldbank.org

Col. 8. The State of the World's Fisheries and Aquaculture, 2008. ftp.fao.org

Col. 9. The State of the World's Fisheries and Aquaculture, 2008. ftp.fao.org

Col. 10. World Development Indicators. The World Bank. (accessed 2010 November 23). worldbank.org

Col 11. Millenium development goal indicators (accessed 2010 November 23). http://unstats.un.org.

Col. 12. World Development Indicators. The World Bank (accessed 2010 November 23). www.worldbank.org

INDEX